Highway Materials, Soils, and Concretes

Second Edition

Harold N. Atkins
Seneca College of Applied Arts
and Technology
Ontario, Canada

Reston Publishing Company, Inc.
A Prentice-Hall Company
Reston, Virginia

Library of Congress Cataloging in Publication Data

Atkins, Harold N.
  Highway materials, soils, and concretes.

  Bibliography: p.
  Includes index.
  1. Road materials. 2. Soils. 3. Concrete.
4. Pavements. I. Title.
TE200.A86   983   628.7'028   82-20526
ISBN 0-8359-2829-2

© 1983 by
Reston Publishing Company, Inc.
A Prentice-Hall Company
Reston, Virginia 22090

10 9 8 7 6 5 4 3 2 1

Printed in the United States of America

Contents

Preface to the Second Edition

The aim of this edition, as of the original, is to provide a practical, clear, concise introduction to soils, aggregates, asphalt, concrete, and pavement construction practices.

The SI system of units has been used throughout, with traditional units also shown. Many of the problems have been repeated in traditional units.

This edition contains additional material in a number of areas, including soil strength testing, geotextiles, asphalt recycling and concrete pavement construction. Specifications of the Canadian Standards Association have also been added. The number of problems has been substantially increased and more photographs have been used to illustrate construction practices.

Comments from students and instructors who have used the book have been both rewarding and valuable aids in the preparation of this revision. In particular I am indebted to Mr. Bill Habkirk of Seneca College and Mr. Barry Munro of the Canadian Portland Cement Association for their suggestions, and to my wife, Anne, for her aid and support.

Harold N. Atkins

1

Engineering Properties of Soils

One of the first steps in civil engineering design is an investigation of soil conditions at the site of the proposed structure. The engineering properties of the soil are important not only as the foundation for the project, but also as a construction material for many structures, including all pavements and earth dams.

1–1 SOIL TYPES

Soil in the engineering field refers to all unconsolidated material in the earth's crust, that is, all material above bedrock. Soil thus includes *mineral particles* (e.g., sand and clay) and *organic material* found in topsoil and marsh deposits, along with the air and water that they contain.

1–1.1 Mineral soil particles result from weathering of the rock that forms the crust of the earth.

Physical weathering–due to the action of frost, water, wind, glaciers, landslides, plant and animal life, and other weathering agents–breaks particles away from the bedrock. These particles are often transported by wind, water, or ice, which both rounds them and further reduces their size. Soils formed through physical weathering are called *granular soils*, the grains or particles of which are similar in nature to the original bedrock.

1

Chemical weathering occurs when water flows through rocks and leaches out some of the mineral components of the rock. New soil particles formed from these minerals are called *clays*. Clay particles are mineral crystals that have very different properties from those of the original bedrock.

1–1.2 The main types of mineral soils are *gravel, sand, silt,* and *clay*. The basic properties of each are indicated in table 1–1.

Table 1–1

BASIC TYPES OF SOIL

Soil	Grain Size	Grain Shape	Soil Group
Gravel	Over 5 mm (over 3/16 in.)	Spherical or cubical	Granular
Sand	From 5 mm (3/16 in.) to smallest visible particles	Spherical or cubical	Granular
Silt	Particles not visible to eye	Spherical or cubical	Granular
Clay	Particles smaller than silt	Flat, plate-shaped grains	Cohesive

Gravel and sand are *coarse-grained* soils, while silt and clay are *fine-grained*. Very large particles are classified as *cobbles* (over 75 mm, or 3 in) or *boulders* (over 200 mm, or 8 in).

Clays are cohesive soils, since the grains are bonded to each other. In nature, clays are not found as separate, unattached grains. A sample, say a handful, would contain millions of grains bonded together. Gravels, sands, and silts, on the other hand, are usually found as individual grains. Moist sands and silts may appear to have some cohesion, since the grains stick to each other. But this "cohesion" is due solely to the moisture film around the grains, and disappears when the soil is dried.

1–1.3 The basic difference in the engineering properties of clays and granular soils arises mainly from the very large variation in size and shape of the grains in the two types of soil.

Because of the clay grain's extremely small size and flat shape, the mass of the grain as a force is negligible when compared to the forces resulting from the surface properties of the grain. Clays have charges on their surface which, owing to the very large surface area per gram of material, govern the behavior of the soil. On account of their chemical composition, clay grains have a surplus of negative charges on their sides and positive charges on their edges. A typical clay grain is illustrated in fig. 1—1.

Two significant results of these surface properties are the water-holding capacity of clays and the structure of clay deposits.

The surface charges attract water molecules, which are held tightly to the surface of the grain. The force holding the water decreases with the water's distance from the particle, but even at a considerable distance there is some holding force. Figure 1—2 illustrates the thickness of the water layer held by two types of clay mineral. (For comparison, it should be noted that a small grain of sand might have a diameter of 0.1 mm, or 100 000 000 pm. A 2-4 pm layer of water on this grain would not be significant.)

FIG. 1—1. Typical clay grain (showing surface charges).

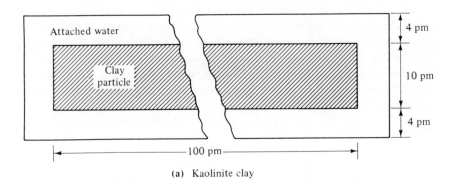

(a) Kaolinite clay

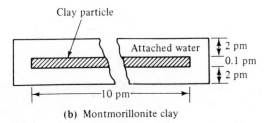

(b) Montmorillonite clay

Note: Not drawn to scale.

FIG. 1—2. Soil moisture on clay particles of typical dimensions— 1 pm = 10^{-9} mm or 10^{-12} m. (Reprinted with permission from Clarkson H. Oglesby, *Highway Engineering*, 3rd ed. New York: Wiley, 1975.)

Figure 1–2 shows that in extremely small-grained clays—such as montmorillonite—the layer of water held to the surface of the grain may be about 4 pm (2 pm on each side). Because the grain is only 0.1 pm thick, the ratio of water volume to mineral solid volume may be as high as 4 to 0.1, or 40 to 1. Of course, many clays are larger than montmorillonite and hold smaller volumes of water relative to the volume of the soil grain. But *all* clays absorb or hold water which remains permanently attached unless conditions change. Clay, when exposed, may dry out due to evaporation. Clay deposits may also lose water when a load, such as a building, is placed on the soil, squeezing the water out. Clay deposits that have dried out may also absorb large quantities of moisture.

This absorption capacity and swelling potential can be visually demonstrated by adding water to a beaker containing about a tablespoon (50 g) of bentonite, a type of montmorillonite clay. If the water and clay are mixed thoroughly, the beaker fills with a wet clay—which is of course mainly water, but appears to be soil solids.

The structure of clay deposits is also governed by the surface properties of the grains. Clays are deposited by settling out in lakes or other bodies of still water. As the grains settle, their surface charges force them to join together in edge-to-side-patterns, since opposite charges attract each other. The resulting structure is therefore very open and flocculent—a "card-house" structure. Granular soils tend to be deposited in a denser configuration, since the force of gravity on the mass of these grains is more important. The structures of clay and granular soil deposits are illustrated in fig. 1–3.

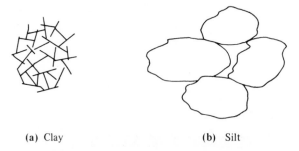

(a) Clay   (b) Silt

FIG. 1–3.   Typical structures of clay and granular soil deposits.

1–1.4   Because of their large grains, sand and gravel particles are readily identified by sight in the field. Organic soils are also easy to classify. With silts and clays, however, individual grains are not visible. The tests described in table 1–2 can be used to identify these soils in the field.

The different reactions in these simple tests can be traced to the difference in grains. Silt grains are coarser and are not bonded tightly together. Therefore

Table 1–2

FIELD TESTS TO IDENTIFY SILTS AND CLAYS

Test	Method	Result
Grittiness	Rub particles between fingers, or taste	Gritty texture–silt; Smooth texture–clay
Toughness	Take a pat of soil, moist enough to be plastic but not sticky, and roll it to a thread about 3 mm (1/8 in) in size in your palm. Fold and reroll thread repeatedly until it crumbles. Lump pieces together and knead to measure toughness.	If the soil is tough or stiff, clay content is high. If it crumbles easily, silt content is high
Shine	Stroke soil with a blade	Dull appearance–silt; Shiny appearance–clay
Dry strength	Allow soil to dry, then squeeze	Powders–silt; Hard to break–clay
Shaking	Squeeze a moistened sample, open hand, then shake or tap your hand	Moisture film comes to surface, glistens–silt; No moisture film–clay

they are gritty, less plastic, and dull when cut. When silt grains are dried, their apparent cohesion disappears, and the sample powders easily. In the shaking test, the saturated silt sample becomes denser when jarred, causing moisture to seep to the surface. This phenomenon is called *dilatancy*. Clay contains grains which are bonded together, and shaking it does not result in an increase in density. Figure 1–4 illustrates results of the shaking test.

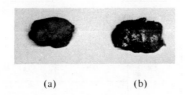

(a)   (b)

FIG. 1–4. Reactions to the shaking test: (a) clay; (b) silt.

1–2 MASS-VOLUME RELATIONS

A soil sample contains (1) mineral–and possibly organic–particles (soil solids), (2) water, and (3) air. The relative amounts of these three *phases* are indications of the soil's properties and condition.

1–2.1 For convenience of calculation, the amounts of air, water, and solids in a soil sample can be represented by a *phase model*, as shown in fig. 1–5.

   The mass and volume of each phase is usually calculated with the aid of a block diagram, as shown in fig. 1–6. The symbols are:

$$V_A = \text{Volume of air}$$

$$V_W = \text{Volume of water}$$

$$V_V = \text{Volume of voids} \ (= V_A + V_W)$$

$$V_D = \text{Volume of dry soil solids}$$

$$V = \text{Total volume}$$

$$M_A = \text{Mass of air} \ (= 0)$$

$$M_W = \text{Mass of water}$$

$$M_D = \text{Mass of dry soil solids}$$

$$M = \text{Total mass}$$

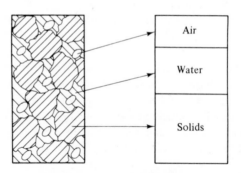

FIG. 1–5. Soil phase model.

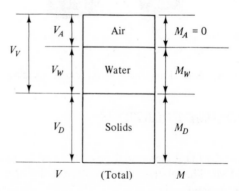

FIG. 1–6. Block diagram for calculation of mass–volume relations.

A soil sample consisting of 10 cm^3 of air, 25 cm^3 of water (mass = 25 g), and 65 cm^3 of soil solids (mass = 175 g) would be represented as shown in example 1–1.

Example 1–1:

$$V_V = 35 \text{ cm}^3 \begin{cases} V_A = 10 \text{ cm}^3 & \text{Air} \\ V_W = 25 \text{ cm}^3 & \text{Water} \quad M_W = 25 \text{ g} \end{cases}$$

$V_D = 65 \text{ cm}^3$ Solids $M_D = 175 \text{ g}$

$V = 100 \text{ cm}^3$ $M = 200 \text{ g}$

1–2.2 To complete the block diagram it is usually necessary to use the relationship between the mass and volume of the water component and of the soil solids component.

For water,

$$\rho_W = M_W / V_W \qquad (1-1)$$

where ρ_W is the density of water. For these calculations density of water is 1 g/cm^3, or 1000 kg/m^3. In example 1–1, ρ_W = 25 g/25 cm^3 = 1 g/cm^3.

For soil solids, or dry soil,

$$\rho_{soil\ solids} = M_D / V_D$$

where $\rho_{soil\ solids}$ is the density of the dry soil solids.

The ratio between the density of the soil solids and the density of water is the *relative density* of the solids. (In traditional terms this quantity is called *specific gravity, G_s*.) Therefore

$$\text{RD (relative density)} = \rho_{soil\ solids} / \rho_W = M_D / (V_D \times \rho_W) \qquad (1-2)$$

or

$$\text{RD} = M_D / (V_D \times \rho_W)$$

In example 1–1

$$\text{RD} = 175 \text{ g}/(65 \text{ cm}^3 \times 1 \text{ g/cm}^3) = 2.69$$

For most soils, relative density is between 2.6 and 2.8.

1-2.3 The properties usually calculated for a soil are:

$$\text{Density } (\rho) \qquad\qquad \rho = \frac{M}{V} \qquad\qquad (1-3)$$

$$\text{Dry density } (\rho_D) \qquad\qquad \rho_D = \frac{M_D}{V} \qquad\qquad (1-4)$$

$$\begin{array}{l}\text{Water content } (w)\\ \text{(moisture content)}\end{array} \qquad w = \frac{M_W}{M_D} \qquad\qquad (1-5)$$

$$\text{Void ratio } (e) \qquad\qquad e = \frac{V_V}{V_D} \qquad\qquad (1-6)$$

$$\text{Degree of saturation } (S) \qquad S = \frac{V_W}{V_V} \qquad\qquad (1-7)$$

$$\text{Porosity } (n) \qquad\qquad n = \frac{V_V}{V} \qquad\qquad (1-8)$$

(In traditional units, the unit weight, γ, is calculated instead of the mass density, ρ. Weight, W, is used in place of mass, M. Following are the formulae replacing 1-3, 1-4, and 1-5 in section 1-2.3.

$$\gamma = \frac{W}{V} \qquad\qquad (1-3a)$$

$$\gamma_D = \frac{W_D}{V} \qquad\qquad (1-4a)$$

$$w = \frac{W_W}{W_D} \qquad\qquad (1-5a)$$

Units for γ and γ_D are usually lb/ft^3. Unit weight of water, γ_w, is 62.4 lb/ft^3.)
In example 1-1, these values are as follows:

$$\text{Density} \qquad\qquad \rho = \frac{M}{V} = \frac{200 \text{ g}}{100 \text{ cm}^3} = 2.00 \text{ g/cm}^3$$

$$\text{Dry density} \qquad\qquad \rho_D = \frac{M_D}{V} = \frac{175 \text{ g}}{100 \text{ cm}^3} = 1.75 \text{ g/cm}^3$$

$$\text{Water content} \qquad w = \frac{M_W}{M_D} = \frac{25 \text{ g}}{175 \text{ g}} \times 100 = 14.3\%$$

Void ratio $\qquad$ $e = \dfrac{V_V}{V_D} = \dfrac{35 \text{ cm}^3}{65 \text{ cm}^3} = 0.54$

Degree of saturation $\qquad$ $S = \dfrac{V_W}{V_V} = \dfrac{25 \text{ cm}^3}{35 \text{ cm}^3} \times 100 = 71\%$

Porosity $\qquad$ $n = \dfrac{V_V}{V} = \dfrac{35 \text{ cm}^3}{100 \text{ cm}^3} \times 100 = 35\%$

1-2.4 To calculate these values, use the following steps:

1. Draw the block diagram and add all known masses and volumes.
2. Calculate other masses and volumes to complete the diagram, using equations (1–1) and (1–2).
3. Substitute in the required formulas [equations (1–3) to (1–8)].

Example 1–2: A soil sample has a volume of 175 cm^3 and a total mass of 300 g. Mass when dried is 230 g. Relative density of the soil solids is 2.70. Find ρ, ρ_D, w, e, S, and n.

1.

Air	
Water	$M_W = 70$ g
Solids	$M_D = 230$ g

$V = 175 \text{ cm}^3 \qquad M = 300 \text{ g}$

$M_W = M - M_D$
$\quad\ \ = 70 \text{ g}$

2.

$V_A = 20 \text{ cm}^3$	Air	
$V_W = 70 \text{ cm}^3$	Water	$M_W = 70$ g
$V_D = 85 \text{ cm}^3$	Solids	$M_D = 230$ g

$V = 175 \text{ cm}^3 \qquad M = 300 \text{ g}$

$V_W = M_W/\rho_W$ [from eq. (1–1)]
$\quad\ = 70\text{g}/(1 \text{ g/cm}^3)$
$\quad\ = 70 \text{ cm}^3$

$V_D = M_D/(RD \times \rho_W)$ [from eq. (1–2)]
$\quad\ = 230 \text{ g}/(2.70 \times 1 \text{ g/cm}^3)$
$\quad\ = 85 \text{ cm}^3$

$V_A = V - (V_D + V_W)$
$\quad\ = 175 \text{ cm}^3 - 155 \text{ cm}^3$
$\quad\ = 20 \text{ cm}^3$

3. $\rho = 300 \text{ g}/175 \text{ cm}^3 = 1.71 \text{ g/cm}^3$

$\rho_D = 230 \text{ g}/175 \text{ cm}^3 = 1.31 \text{ g/cm}^3$

$w = 70 \text{ g}/230 \text{ g} = 30.4\%$

$e = 90 \text{ cm}^3/85 \text{ cm}^3 = 1.06$

$S = 70 \text{ cm}^3/90 \text{ cm}^3 = 78\%$

$n = 90 \text{ cm}^3/175 \text{ cm}^3 = 51\%$

Example 1–3: Given that $W_D = 53.71$ lb, $W_W = 14.91$ lb, $S = 100\%$, and RD = 2.65, find $\rho, w, e.$

$$S = 100\%$$

$$\therefore V_A = 0$$

$$\left(S = \frac{V_W}{V_W + V_A} = 1.0 \right)$$

$V_W = 0.239 \text{ ft}^3$	Water	$W_W = 14.91$ lb
$V_D = 0.325 \text{ ft}^3$	Solids	$W_D = 53.71$ lb
$V = 0.564 \text{ ft}^3$		$W = 68.62$ lb

$$V_D = \frac{W_D}{RD \times \gamma_W} = \frac{53.71 \text{ lb}}{2.65 \times 62.4 \text{ lb/ft}^3} = 0.325 \text{ ft}^3$$

$$V_W = \frac{W_W}{\gamma_W} = \frac{14.91 \text{ lb}}{62.4 \text{ lb/ft}^3} = 0.239 \text{ ft}^3$$

Therefore:

$$\gamma = \frac{68.62 \text{ lb}}{0.564 \text{ ft}^3} = 122 \text{ lb/ft}^3$$

$$w = \frac{14.91 \text{ lb}}{53.71 \text{ lb}} = 27.8\%$$

$$e = \frac{0.239 \text{ ft}^3}{0.325 \text{ ft}^3} = 0.74$$

1-2.5 Problems in which the density is given should be solved by assuming a total volume of 1 cm^3 or 1 m^3 or 1 ft^3.

Example 1-4: The density of a soil is 1900 kg/m^3, the dry density is 1650 kg/m^3, and RD is 2.72. Find w and e. Assume $V = 1$ m^3. Therefore $M = 1900$ kg, $M_D = 1650$ kg, and $M_W = 250$ kg.

$$V_D = \frac{1650 \text{ kg}}{2.72 \times 1000 \text{ kg/m}^3} = 0.607 \text{ m}^3$$

$$V_W = \frac{250 \text{ kg}}{1000 \text{ kg/m}^3} = 0.250 \text{ m}^3$$

$$V_A = 1 - (0.607 + 0.250) = 0.143 \text{ m}^3$$

$V_A = 0.143$ m^3	Air	
$V_W = 0.250$ m^3	Water	$M_W = 250$ kg
$V_D = 0.607$ m^3	Solids	$M_D = 1650$ kg
$V = 1$ m^3		$M = 1900$ kg

Therefore:

$$w = \frac{250 \text{ kg}}{1650 \text{ kg}} = 15.2\%$$

$$e = \frac{0.393 \text{ m}^3}{0.607 \text{ m}^3} = 0.65$$

1-2.6 Many problems give the water content, along with total density or total mass. The water content measures the mass of water as a percentage of the mass of dry soil solids, not as a percentage of the total mass. Therefore the following relationship must be used.

$$M_D = \frac{M}{1 + w} \quad \text{or} \quad \rho_D = \frac{\rho}{1 + w} \tag{1-9}$$

(*Note:* w is expressed as a ratio, not as a percentage.)

Example 1–5: The total mass of a sample is 51.5 g, the total volume is 28.3 cm³, the water content is 16.5%, and the RD is 2.70. Find ρ, ρ_D, and e.

V_A = 4.6 cm³ | Air
V_W = 7.3 cm³ | Water | M_W = 7.3 g
V_D = 16.4 cm³ | Solids | M_D = 44.2 g
V = 28.3 cm³ | | M = 51.5 g

$$M_D = \frac{51.5 \text{ g}}{1 + 0.165} = \frac{51.5}{1.165} = 44.2 \text{ g}$$

$$M_W = 51.5 - 44.2 = 7.3 \text{ g}$$

(*Check:* 7.3/44.2 = 16.5%)

$$V_D = \frac{44.2 \text{ g}}{2.70 \times 1 \text{ g/cm}^3} = 16.4 \text{ cm}^3$$

$$V_W = \frac{7.3 \text{ g}}{1 \text{ g/cm}^3} = 7.3 \text{ cm}^3$$

$$V_A = 28.3 - (16.4 + 7.3) = 4.6 \text{ cm}^3$$

Therefore:

$$\rho = 51.5 \text{ g}/28.3 \text{ cm}^3 = 1.82 \text{ g/cm}^3$$

$$\rho_D = 44.2 \text{ g}/28.3 \text{ cm}^3 = 1.56 \text{ g/cm}^3$$

$$e = 11.9 \text{ cm}^3/16.4 \text{ cm}^3 = 0.73$$

1–2.7 Problems in which the void ratio and relative density are the main known values are usually solved by assuming a one-unit volume of soil solids.

Example 1–6:   Given that e = 1.24, RD = 2.71, and w = 42%, find ρ, ρ_D, and S. Assume V_D = 1 m³. Since e = 1.24 = V_V/V_D, V_V = 1.24 m³.

$V_V = 1.24\ m^3 \begin{cases} V_A = 0.102\ m^3 \\ V_W = 1.138\ m^3 \end{cases}$

Air	
Water	$M_W = 1138$ kg

$V_D = 1\ m^3$ Solids $M_D = 2710$ kg

$V = 2.24\ m^3$ $M = 3848$ kg

$$M_D = V_D \times RD \times \rho_W \ \text{[from eq. (1–2)]}$$
$$= 1\ m^3 \times 2.71 \times 1000\ kg/m^3$$
$$= 2710\ kg$$
$$M_W = w \times M_D = 0.42 \times 2710\ kg$$
$$= 1138\ kg$$
$$V_W = M_W/\rho_W = \frac{1138\ kg}{1000\ kg/m^3} = 1.138\ m^3$$
$$V_A = 1.24 - 1.138 = 0.102\ m^3$$
$$V = 0.102 + 1.138 + 1.00 = 2.24\ m^3$$

Therefore:

$$\rho = \frac{3848\ kg}{2.24\ m^3} = 1720\ kg/m^3$$

$$\rho_D = \frac{2710\ kg}{2.24\ m^3} = 1210\ kg/m^3$$

$$S = \frac{1.138\ m^3}{1.24\ m^3} = 92\%$$

Example 1–7: Assuming an average relative density of soil solids of 2.65, find the dry density of saturated sample at a water content of 20%. Assume $V_D = 1\ m^3$. $S = 100\%$; therefore $V_A = 0$.

$$M_D = V_D \times RD \times 1000\ kg/m^3$$
$$= 1\ m^3 \times 2.65 \times 1000\ kg/m^3$$

$$= 2650 \text{ kg}$$

$$M_W = 0.20 \times M_D = 0.20 \times 2650$$

$$= 530 \text{ kg}$$

$$V_W = M_W / \rho_W = 530 \text{ kg}/1000 \text{ kg/m}^3$$

$$= 0.530 \text{ m}^3$$

Therefore:

$V_W = 0.530 \text{ m}^3$	Water	$M_W = 530 \text{ kg}$
$V_D = 1 \text{ m}^3$	Solids	$M_D = 2650 \text{ kg}$

$$V = 1.530 \text{ m}^3$$

$$\rho_D = \frac{2650 \text{ kg}}{1.530 \text{ m}^3} = 1730 \text{ kg/m}^3$$

1–2.8 Typical values for void ratio, density, and water content of soils found in nature are shown in table 1–3.

1–2.9 The condition of certain soils in field deposits is often described as loose or dense, or in similiar terms indicating the soil's density in relation to the maximum and minimum possible densities for that type of soil.

This condition is especially important in assessing the stability of granular soils.

Dry density of the soil is determined in the field. A sample is tested in the laboratory to find its maximum possible dry density (compacted in a vibrating mold at an optimum water content) and its minimum dry density (poured slowly into a mold without any vibration or water). The *density index* (I_D) is then found as follows:

$$I_D = \frac{\rho_{D \text{ max}}}{\rho_D} \times \frac{\rho_D - \rho_{D \text{ min}}}{\rho_{D \text{ max}} - \rho_{D \text{ min}}} \qquad (1-10)$$

(In traditional units, this property has been termed *relative density*.)

Table 1-3
TYPICAL VOID RATIO, DENSITY, AND WATER CONTENT VALUES OF SOILS

Soil	Void Ratio	Density (kg/m³)		Unit Weight (lb/ft³)		Water Content (%)
		Dry	Saturated	Dry	Saturated	Saturated
Gravel-sand mixture	0.5	1800	2100	110	135	20
Uniform sand	0.7	1600	2000	100	125	25
Hard glacial till	0.4	1900	2200	120	140	15
Hard glacial clay	0.6	1700	2100	105	130	20
Soft clay	2.0	1400	1600	80	100	75
Soft montmorillonite-type clay	5.0	500	1300	30	80	180

Example 1–8: Dry density of a soil in the field is 1730 kg/m^3. In laboratory tests, maximum dry density was found to be 1870 kg/m^3 and minimum dry density, 1420 kg/m^3.

$$I_D = \frac{1870}{1730} \times \frac{1730 - 1420}{1870 - 1420} = 0.74 \text{ or } 74\%$$

The density index can also be found using the void ratio of the soil and the maximum and minimum void ratios as found in the laboratory tests described above.

$$I_D = \frac{e_{max} - e}{e_{max} - e_{min}} \qquad (1-11)$$

1–3 CLASSIFICATION TESTS

The two most important types of tests used in classifying soils are:

— *Grain size*, to measure grain sizes; and
— *Plasticity*, to measure grain types.

1–3.1 Results of grain size tests are plotted on a *grain size distribution graph*, as shown in fig. 1–7. The grain size is shown on a logarithmic scale along the bottom of the graph. The percentage by mass of a soil that is smaller than (or "passes") each size is shown on the left side. Line *A* on the graph represents a soil in which

— 100% of a sample is smaller than 10 mm
— 78% of a sample is smaller than 1 mm
— 36% of a sample is smaller than 0.1 mm
— 16% of a sample is smaller than 0.01 mm.

The smooth curve joining these points is known as the *grain size distribution curve*. This curve rarely intersects the 0% line, as the size of the smallest grain in a sample is seldom known.

1–3.2 Grain sizes in soil samples are found by means of two tests. The *sieve analysis* is used for sands and gravels; the *hydrometer test*, for silts and clays. If significant quantities of both coarse- and fine-grained soils are in the sample, the results of both tests may have to be combined to plot the grain size distribution curve.

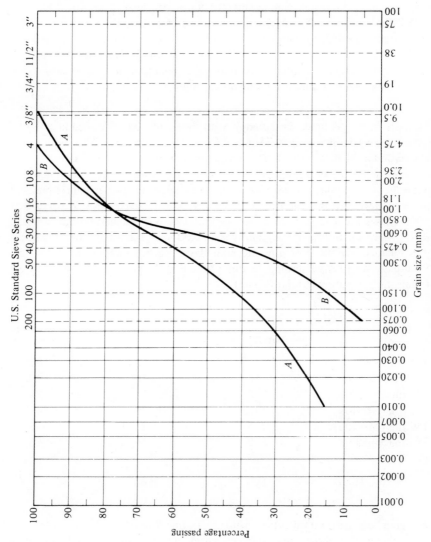

FIG. 1–7. Graph for grain size distribution curve.

17

1-3.3 Standard sieves are sized according to the space between each pair of wires. Traditionally sizes have been specified in inches for sieves that are one-fourth inch and larger and by numbers for smaller sizes. A No. 4 sieve is approximately 5 mm (3/16 in) in size and, as the size decreases, the numbers increase as they originally indicated the number of wires per inch. The No. 200 sieve (0.0029 in) is the smallest in common use. In current practice, sieves are usually designated according to the opening in millimeters or micrometers, for example, 4.75 mm for the No. 4 sieve and 0.075 mm or 75 μm for No. 200. See fig. 1—7 for the opening sizes (in millimeters) that correspond to various sieve sizes.

Grain size distribution is calculated as follows.

Example 1—9: Assume that a sample with a mass of 500 g is placed in a nest of sieves ranging in size from 4.75 mm (No. 4) to 75 μm (No. 200), with a pan placed under the nest to catch all grains passing through the 75 μm sieve.

We shake the sieves to separate the grains according to size, and then determine the mass retained in each sieve. We then calculate the percentage retained on each sieve and the cumulative percentage passing each sieve. We find that 100% passes the 4.75 mm sieve. We see that 10% was retained on the 2.00 mm sieve meaning that 90% of the grains passed 2.00 mm (100 - 10). Fifteen percent was retained on the 850 μm, meaning that 75% passed 850 μm (90 - 15). Eventually we find 5% passing 75 μm. (Note that the percentage passing 75 μm equals the percentage left in the pan.)

Sieve	*Mass Retained*	*% Retained*	*Cumulative % Passing*
4.75 mm (No. 4)	0 g	0%	100%
2.00 mm (No. 10)	50	10	90
850 μm (No. 20)	75	15	75
425 μm (No. 40)	175	35	40
150 μm (No. 100)	125	25	15
75 μm (No. 200)	50	10	5
Pan	25	5	
Totals	500	100	

The grain size distribution curve for this soil is line *B* in fig. 1—7.

Figure 1—8 shows some standard sieves and the grains retained on them in a sieve analysis.

1—3.4 The hydrometer test is used to find the size of smaller grains. It is a sedimentation test: the rate at which particles settle is used as an indication of their size.

FIG. 1–8. Grain size determination by sieve analysis. Sieve sizes (from left): top row—4.75 mm (No. 4), 1.18 mm (No. 16), 300 μm (No. 50); bottom row—75 μm (No. 200), Pan.

Stokes' law states that particles in a suspension settle out at a rate which varies with their size. Using this law, we can calculate the size of particle that has settled a known distance in the suspension at any time from the beginning of sedimentation. A hydrometer is used to measure the density of the water—soil suspension at various times as the grains settle. The size of the particle that has settled to the center of the hydrometer bulb can be calculated, and the density of the solution indicates the percentage of the sample still in the suspension. Using this data, a grain size distribution curve can be established.

Figure 1–9 shows how the depth to the center of the hydrometer changes as particles settle out over a period of time.

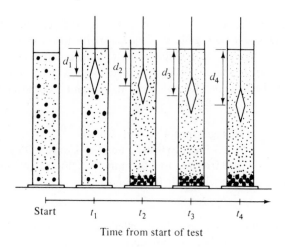

FIG. 1–9. The hydrometer test.

1–3.5   Following are some of the ways in which the grain size distribution curve is used to help describe and classify a soil:

1. *Shape.* A soil composed mainly of one size, such as a beach sand, is called a *uniform* soil, and has a grain size distribution curve such as curve *A* (fig. 1–10). A soil that contains a wide range of grain sizes, such as soil *B*, is termed *well-graded.*

2. *Effective Size.* Since the amount and type of fine grains ("fines") in a soil are very important in assessing the properties of that soil, the effective size is taken as the 10% size, that is, the grain size that only 10% of the grains are finer than. This is obtained from reading the grain size on the curve where the 10% passing line is intersected. For Soil *B*, the effective size is 0.09 mm.

3. *Uniformity Coefficient (C_u).* This value gives some indication of the shape of the curve and the range of particle sizes that a soil contains, especially in the more important fine part of the soil. The formula is:

$$C_u = D_{60}/D_{10} \qquad (1-12)$$

For Soil *B* $C_u = 7.0 \text{ mm}/0.09 \text{ mm} = 78$

4. *Coefficient of Curvature (C_z).* This is another measurement of the shape of the curve. It is calculated as follows:

$$C_z = (D_{30})^2/(D_{60} \times D_{10}) \qquad (1-13)$$

For Soil *B* $C_z = 1.1^2/(7 \times 0.09) = 1.9$

5. *Textural Classification.* This type of soil classification is based on grain size. Other systems use both grain size and plasticity to classify soils, since grain size is not very accurate as a sole criteria in differentiating between silt and clay. However, textural classification systems are widely used. The American Society for Testing and Materials (ASTM) system is as follows:

Gravel	larger than 4.75 mm (No. 4)
Sand	4.75 mm to 0.075 mm (No. 4 to No. 200)
Silt	0.075 mm to 0.005 mm (No. 200 to .005 mm)
Clay	smaller than 0.005 mm

Note: This applies to soil particles smaller than 75 mm (3 in). Particles over 75 mm are classified as cobbles or boulders.

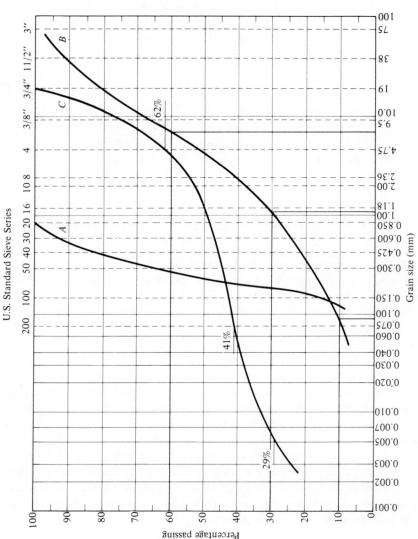

FIG. 1–10.   Grain size distribution curves.

Example 1–10: For example, curve *C* on the grain size distribution in fig. 1–8 shows the following:

> 100% passes 19 mm sieve
> 62% passes 4.75 mm sieve
> 41% passes 75 μm sieve
> 29% passes 0.005-mm size

Therefore:

38% of the grains are gravel sizes	(100% – 62%)
21% of the grains are sand sizes	(62% – 41%)
12% of the grains are silt sizes	(41% – 29%)
29% of the grains are clay sizes	(29% passes 0.005)
100% (total)	

Other methods of classifying soils by grain size are shown in fig. 1–11.

1–3.6 The other main test used in classifying soils is the *plasticity test*. This test measures the amount of water that a soil absorbs, or that a soil requires before it will roll like a plastic material and act like a fluid material.

A dry clay soil is hard, brittle, and cracked ($w = 0$). If this soil is ground up and water is added, it becomes more plastic. When enough water is added that it is possible to roll the soil to a thread, it is at its *plastic limit*. That is, if the water content required to roll the soil is 27%, then the plastic limit (w_P) for this soil is 27. If more water is added, the soil becomes softer and more like a liquid. At a water content when it will flow in a *liquid limit cup*, it is at its *liquid limit* (w_L). If this moisture content is 43%, the liquid limit for this soil is 43. The range of water contents over which this soil is plastic is called the *index of plasticity* ($I_P = w_L - w_P$). In this case, $I_P = 43 - 27 = 16$.

A soil that is dried to below its plastic limit shrinks as water is removed, but retains its shape until it reaches a certain point called the *shrinkage limit* (w_S). Further drying results in cracking of the soil, as the loss in volume due to removal of water cannot be accomplished by grains coming closer together. The limits are illustrated on this chart:

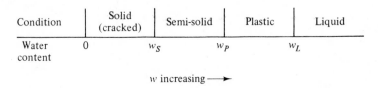

Condition	Solid (cracked)	Semi-solid	Plastic	Liquid
Water content	0 w_S		w_P	w_L

w increasing ⟶

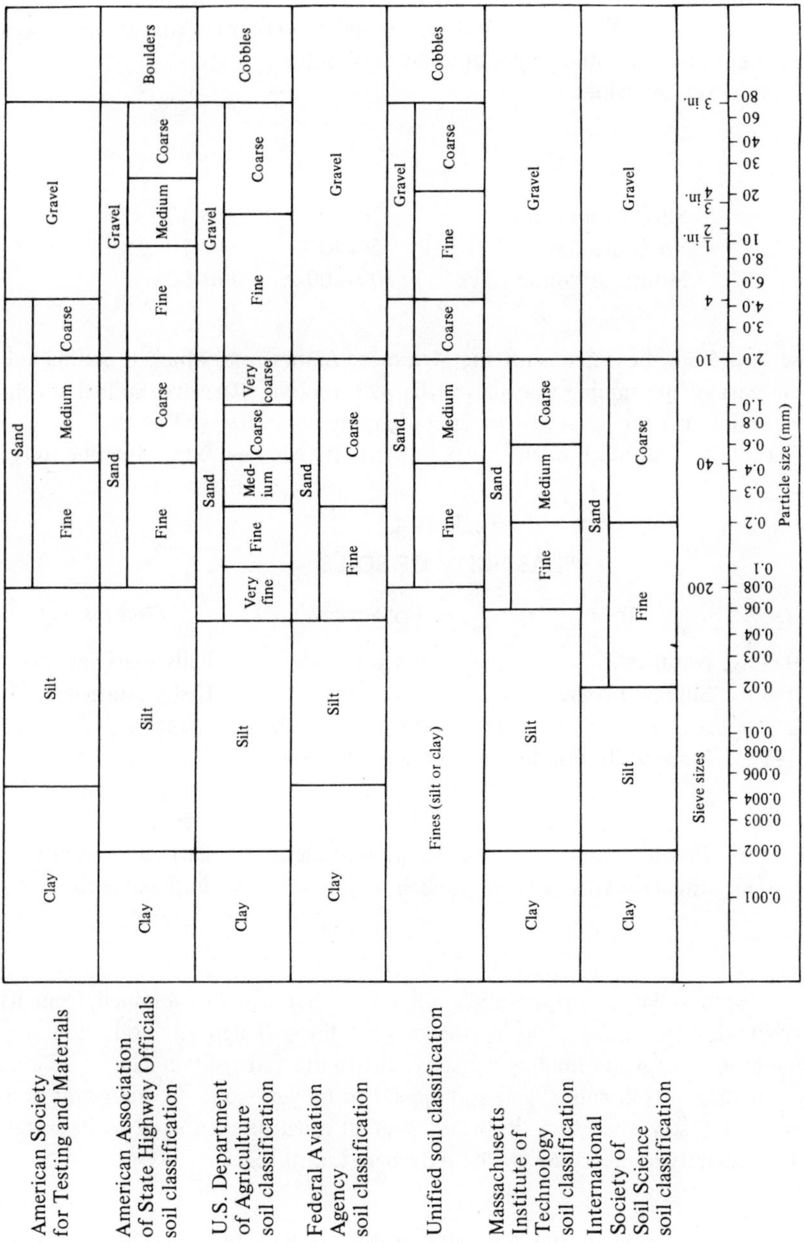

Classification	Fine fraction		Sand					Gravel			Largest
American Society for Testing and Materials	Clay	Silt	Fine	Medium		Coarse		Gravel			Boulders
American Association of State Highway Officials soil classification	Clay	Silt	Fine			Coarse		Fine	Medium	Coarse	Cobbles
U.S. Department of Agriculture soil classification	Clay	Silt	Very fine	Fine	Medium	Coarse	Very coarse	Fine		Coarse	
Federal Aviation Agency soil classification	Clay	Silt	Fine			Coarse		Gravel			
Unified soil classification	Fines (silt or clay)		Fine	Medium		Coarse		Fine		Coarse	Cobbles
Massachusetts Institute of Technology soil classification	Clay	Silt	Fine	Medium		Coarse		Gravel			
International Society of Soil Science soil classification	Clay	Silt	Fine			Coarse		Gravel			

Sieve sizes: 200 40 10 4 ¾ in 3 in

Particle size (mm): 0.001 0.002 0.003 0.004 0.006 0.008 0.01 0.02 0.03 0.04 0.06 0.08 0.1 0.2 0.3 0.4 0.6 0.8 1.0 2.0 3.0 4.0 6.0 8.0 10 20 30 40 60 80

FIG. 1–11. Soil types based on grain size.

The plasticity of a soil gives a very good indication of the types of soil grain in the sample. Low values of the limits are found in silty soil, with higher values in clay soils. Very high limits are found in extremely fine-grained clays such as bentonite and other types of montmorillonite.

Some typical values are:

	w_P	w_L
Silt-clay mixtures	20–30	25–40
Kaolinite clays	20–40	40–70
Montmorillonite clays	100–200	300–600

These limits are measured with the *Atterberg Limits test*, which is conducted on the part of the sample that passes the 425 μm (No. 40) sieve. Soils that cannot be rolled to a thread at any water content are *nonplastic* (NP).

Table 1–4 shows how the index of plasticity can be used to describe soils.

Table 1–4

PLASTICITY OF SOILS

I_P	Term	Dry Strength	Field Test
0–3	Nonplastic	Very low	Falls apart easily
4–6	Slightly plastic	Low	Easily crushed by fingers
7–15	Moderately plastic	Low to medium	Slight pressure required to crush
16–35	Plastic	Medium to high	Difficult to crush
Over 35	Highly plastic	High	Impossible to crush

A second important property of a soil that can be obtained from its Atterberg Limits values is the consistency of the soil deposit. Soils vary considerably in their water-holding capacity. Bentonite can contain a larger volume of water than of soil solids, and still appear to be very solid. To judge the consistency of a soil, therefore, its water content must be compared to its water-holding capacity as measured by the Atterberg Limits test.

Example 1–11: Soil *A* has a natural water content of 35%; Soil *B*, 45%. Atterberg Limits tests are conducted on samples of these soils, with the following results:

Soil A $w_P = 25$ $w_L = 32$
Soil B $w_P = 40$ $w_L = 80$

Describe and compare the consistencies of the two soils.

Soil A is very soft. Its natural water content is above its liquid limit. Soil B is quite firm, as its natural water content is just slightly above its plastic limit and well below its liquid limit. Therefore Soil B is much harder and firmer than Soil A, *even though it contains more water.* The location of Soils A and B in their natural state on the plasticity line can be shown as follows:

Condition	Solid	Plastic	Liquid
Water content 0		$w_P \uparrow$	$w_L \uparrow$
		Soil B	Soil A

The Atterberg Limits test for the plastic and liquid limits of a soil is illustrated in fig. 1–12. A sample of moist soil is rolled to a thread on a glass plate, formed into a ball, and re-rolled. Water is being removed as this happens, and at a certain water content the sample crumbles. This water content is the plastic limit. To obtain the liquid limit, the sample is placed in the cup as shown, and a groove is made through its center. The cup is jarred to make the soil flow to close the groove. When the groove closes under specified conditions, the soil is at its liquid limit.

FIG. 1–12. Atterberg Limits Test.

1−4   SOIL CLASSIFICATION

For engineering purposes, soils are frequently classified into groups. Two common classification systems are (1) the *Unified Soil Classification System*, used for general engineering purposes and published as ASTM Standard D 2487, and (2) the *AASHTO System*, developed by the American Association of State Highway and Transportation Officials and often used for soils in highway engineering. Other systems have been developed, but these two are the most commonly used systems in North America.

1−4.1   In the Unified System, soils are usually given a two-letter designation. The first letter indicates the main soil type, and the second modifies the first. The symbols are:

	Symbol	Description
	G	Gravel
	S	Sand
1st letter	M	Silt
	C	Clay
	O	Organic
	Pt	Peat
	W	Well-graded
	P	Poorly graded
2nd letter	M	Silty fines
	C	Clayey fines
	H	High plasticity
	L	Low plasticity

Soils are divided into three general areas: (1) *coarse-grained* soils, including gravels (G) and sands (S), where the second letter indicates gradation (W, P) or type of fines (M, C); (2) *fine-grained* soils, including silts (M), clays (C), and organic soils (O) (depending on plasticity), where the second letter indicates high (H) or low (L) plasticity; and (3) *peaty* soils (Pt), which contain a large proportion of fibrous organic matter.

Figure 1−13 gives the Unified System's classification system. The grain size distribution and the Atterberg Limits test results are required. Figure 1−14 indicates the main steps in the classification procedure, and the following examples illustrate its use.

Example 1–12: Grain size distribution:

Passes 38 mm (1½-in.)	100%
19 mm (3/4-in.)	90%
9.5 mm (3/8-in.)	77%
4.75 mm (No. 4)	53%
425 μm (No. 40)	33%
75 μm (No. 200)	20%

Atterberg Limits: $w_L = 48$, $w_p = 31$, $I_p = 17$.

The sample contains 80% (100% - 20%) coarse-grained sizes. 47% of these grains are retained on 4.75 mm, and 33% pass 4.75 mm. Therefore the soil is a gravel (G). Over 12% passes 75 μm. The type of fines is silty. (w_L and I_p plot below line *A*.) Therefore the soil classification is GM.

Example 1–13: Grain size distribution:

Passes 9.5 mm (3/8-in.)	100%
4.75 mm (No. 4)	60%
425 μm (No. 40)	30%
150 μm (No. 100)	10%
75 μm (No. 200)	4%

The sample contains 96% (100% - 4%) coarse-grained sizes. 40% of these grains are retained on 4.75 mm and 56% pass 4.75 mm. Therefore the soil is a sand (S). Less than 5% pass 75 μm. Therefore the soil is a clean sand, and the shape of the gradation curve must be found. (Note, in this example the D_{60}, D_{30} and D_{10} sizes correspond to 4.75 mm, 425 μm and 150 μm sieves. Usually the grain size distribution curve would have to be plotted.)

$$C_u = D_{60}/D_{10} = 4.75/0.150 = 32$$
$$C_z = (D_{30})^2/(D_{60} \times D_{10}) = (0.425)^2/(4.75 \times 0.150) = 0.25$$

This sample does not meet both requirements for a well-graded sand, so the classification is SP.

Unified Soil Classification System
(ASTM designation D-2487)

Major Division			Group Symbols	Typical Names	Classification Criteria
Coarse-grained soils — More than 50% retained on 75 μm (No. 200) sieve	Gravels — 50% or more of coarse fraction retained on 4.75 mm (No. 4) sieve	Clean gravels	GW	Well-graded gravels and gravel-sand mixtures, little or no fines	$C_u = D_{60}/D_{10}$ Greater than 4 $C_z = \dfrac{(D_{30})^2}{D_{10} \times D_{60}}$ Between 1 and 3
			GP	Poorly graded gravels and gravel-sand mixtures, little or no fines	Not meeting both criteria for GW
		Gravels with fines	GM	Silty gravels, gravel-sand-silt mixtures	Atterberg limits plot below "A" line or plasticity index less than 4
			GC	Clayey gravels, gravel-sand-clay mixtures	Atterberg limits plot above "A" line and plasticity index greater than 7
	Sands — More than 50% of coarse fraction passes 4.75 mm (No. 4) sieve	Clean sands	SW	Well-graded sands and gravelly sands, little or no fines	$C_u = D_{60}/D_{10}$ Greater than 6 $C_z = \dfrac{(D_{30})^2}{D_{10} \times D_{60}}$ Between 1 and 3
			SP	Poorly graded sands and gravelly sands, little or no fines	Not meeting both criteria for SW
		Sands with fines	SM	Silty sands, sand-silt mixtures	Atterberg limits plot below "A" line or plasticity index less than 4
			SC	Clayey sands, sand-clay mixtures	Atterberg limits plot above "A" line and plasticity index greater than 7

Classification on basis of percentage of fines

GW, GP, SW, SP — Less than 5% pass 75 μm sieve

GM, GC, SM, SC — More than 12% pass 75 μm sieve

Borderline classification requiring use of dual symbols — 5% to 12% pass 75 μm sieve

Plasticity chart for the classification of fine-grained soils.

	Symbol	Typical names
Fine-grained soils 50% or more passes 75 μm (No. 200) sieve — **Silts and Clays** Liquid limit 50% or less	ML	Inorganic silts, very fine sands, rock flour, silty or clayey fine sands
	CL	Inorganic clays of low to medium plasticity, gravelly clays, sandy clays, silty clays, lean clays
	OL	Organic silts and organic silty clays of low plasticity
Silts and Clays Liquid limit greater than 50%	MH	Inorganic silts, micaceous or diatomaceous fine sands or silts, elastic silts
	CH	Inorganic clays of high plasticity, fat clays
	OH	Organic clays of medium to high plasticity
Highly organic soils	Pt	Peat, muck and other highly organic soils — Fibrous organic matter; will char, burn, or glow

FIG. 1–13. Unified Soil Classification System.

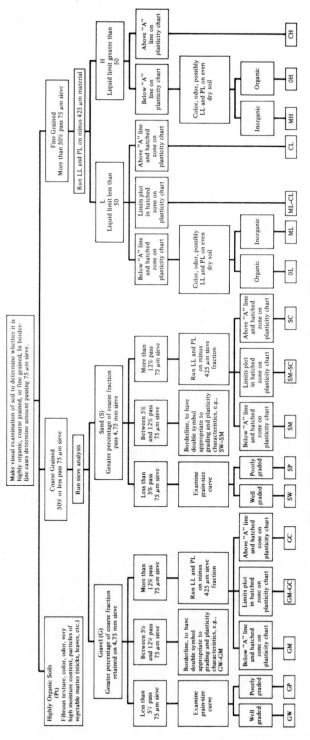

FIG. 1–14. Classification of soils by the Unified System.

Example 1–14: Grain size distribution:

Passes 4.75 mm (No. 4)	88%	60% size–2.0 mm
425 μm (No. 40)	28%	30% size–0.5 mm
75 μm (No. 200)	9%	10% size–0.08 mm

Atterberg Limits: $w_P = 20$, $w_L = 31$, and $I_P = 11$.
   This sample is:

1. Coarse-grained, since 91% is retained on 75 μm.
2. Sand, since 12% is retained on 4.75 mm and 79% passes 4.75 mm.
3. Mixed, since 5–12% passes 75 μm.

w_L and I_P plot above line A; therefore, fines are of the clay type (C).

$$C_u = 2.0/0.08 = 25$$
$$C_z = (0.5)^2/(2.0 \times 0.08) = 1.6 \quad \therefore \text{ well-graded (W)}$$

The classification is SW–SC.

Example 1–15: Grain size distribution:

Passes 75 μm (No. 200)   80%

Atterberg Limits: $W_L = 65$ and $I_p = 42$.
   Therefore this is a fine-grained soil, above line A, high in compressibility.
The classification is CH.

Example 1–16: Grain size distribution:

Passes 75 μm (No. 200)   100%

Atterberg Limits: $w_L = 20$ and $I_p = 5$.
   Therefore, this is a fine-grained soil, mixed type; the limits plot is shown
in the hatched zone of the plasticity chart in fig. 1–13.
   The classification is CL–ML.

1–4.2 The ASSHTO system groups soils into seven main groups—A-1 to A-7—
based generally on the desirability of the soil as a subgrade for highway construc-

tion. Again, grain size distribution and plasticity values are the criteria to classify soils. The AASHTO classification system is shown in table 1–5.

Given the grain size and plasticity data, you must check each classification—starting from the left. The first group that the test data fit is the correct one. A-3 soils are listed to the left of the A-2 soils to accommodate this left-to-right elimination system, not because they are superior as subgrade material.

The A-1 soils are gravels and coarse sands with few fines and low plasticity. A-3 contains clean, fine sands. A-2 soils are granular soils with up to 35% fines. Subgroups A-2-4 and A-2-5 are gravels or sands that contain either excessive amounts of fines or fines with too high a plasticity to fit into A-1. Subgroups A-2-6 and A-2-7 contain more plastic or clayey fines. A-4 and A-5 are silty soils. A-6 and A-7 are clayey soils.

The following examples illustrate the use of the AASHTO system.

Example 1–17: Soil sample test results:

Passes 38 mm (1 1/2-in)	100%
2.00 mm (No. 10)	65%
425 μm (No. 40)	45%
75 μm (No. 200)	30%

$$w_L = 35$$
$$I_P = 21$$

The sample cannot be classified as A-1-a (over 50% passes 2.00 mm); A-1-b (over 25% passes 75 μm); A-3 (less than 51% passes 425 μm); A-2-4 (I_P is greater than 10); or A-2-5 (w_L is less than 41). But the sample does meet the requirements of the A-2-6 classification.

Example 1–18: Soil sample test results:

Passes 4.75 mm (No. 4)	100%
425 μm (No. 40)	73%
75 μm (No. 200)	65%

$$w_L = 63$$
$$I_P = 41$$

The sample cannot be classified in any of the sand and gravel classes, since more than 35% passes 75 μm. Plasticity characteristics indicate that the sample is in the A-7 group, which contains two subgroups. Liquid limit (63) minus 30 = 33. Since the plasticity index (41) is greater than 33, the classification is A-7-6.

Table 1-5
AASHTO SOIL CLASSIFICATION SYSTEM
(AASHTO STANDARD M145)*

General Classification	Granular Materials (35% or less passing No. 200, 75 μm)							Silt-Clay Materials (more than 35% passing No. 200, 75 μm)			
Group Classification	A-1-a	A-1-b	A-3	A-2-4	A-2-5	A-2-6	A-2-7	A-4	A-5	A-6	A-7-5, A-7-6
Sieve analysis, % passing:											
No. 10 (2.00 mm)	50 max	......	......	......	......	......	......				
No. 40 (425 μm)	30 max	50 max	51 min	......	......	......	......				
No. 200 (75 μm)	15 max	25 max	10 max	35 max	35 max	35 max	35 max	36 min	36 min	36 min	36 min
Characteristics of fraction passing No. 40 (425 μm):											
Liquid limit	......	......	......	40 max	41 min	40 max	41 min	40 max	41 min	40 max	41 min
Plasticity index	6 max	6 max	N.P.	10 max	10 max	11 min	11 min	10 max	10 max	11 min	11 min†
Usual types of significant constituent materials	Stone fragments, gravel and sand		Fine sand	Silty or clayey gravel and sand				Silty soils		Clayey soils	
General rating as subgrade	Excellent to good							Fair to poor			

* Reprinted by permission of the American Association of State Highway and Transportation Officials from *AASHTO Materials*, 12th edition, 1978.
† Plasticity index of A-7-5 subgroup is equal to or less than LL minus 30. Plasticity index of A-7-6 subgroup is greater than LL minus 30.

1–5 SOIL WATER

Soil is made up of soil particles, water, and air. This section discusses the types of water in soils, their location, the forces governing their movement, and tests involved in flow measurement.

1–5.1 The types of water found in soil, as illustrated in fig. 1–15, may be classified as follows:

1. *Free water* or *gravitational water,* which is found below the groundwater table and is free to flow under the laws of gravity.
2. *Capillary water,* which is brought up through the soil pores (the spaces between soil grains) above the groundwater table due to surface tension forces.
3. *Attached water* or *held water,* which is the water in the moisture film around soil grains.

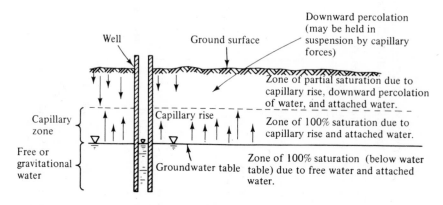

FIG. 1–15. Types of water in soils.

The location and movement of free water and capillary water are discussed below. Attached water acts as part of the soil grain and does not move unless forced out by loading. The quantity of attached water may be very large in clays, due to the fact that the moisture film is relatively thick with respect to the thickness of the soil grain.

1–5.2 The *groundwater table* is the surface below which all soil pores are filled with water which is free to flow. It is the surface at which the pressure in the water is atmospheric. Below this surface the water pressure increases, as it does below the surface of any body of water. Soil is usually fully saturated below the water table. The table rises and falls, depending on (1) the climate, rising in wet seasons (as rain adds to the quantity) and falling in dry seasons; (2) man-made changes, such as pumping; and (3) changes in the elevation of lakes and streams.

At lakes, streams, and swamps, the water table is usually above the ground surface, at the water surface. Springs indicate where the water table appears at the ground surface. Since water flows relatively quickly through coarse-grained soils, the water table is usually found close to the governing lake or stream elevation. In clays, water flows extremely slowly, if at all, and the water table is usually near the surface, lowered only by evaporation near the surface.

A *perched water table* is one that is located above the true water table. It results from water's being trapped above an impermeable layer.

Artesian water is water under pressure that is prevented from flowing by an impermeable layer.

Water table locations are shown in fig. 1–16.

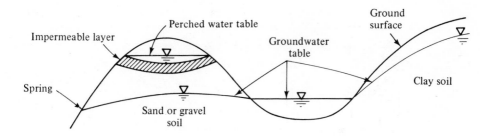

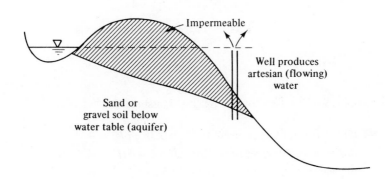

FIG. 1–16. Groundwater table.

1–5.3   Rate of water flow through soils depends on the *permeability* of the soil, which is defined by Darcy's law (see fig. 1–17):

$$q = k i A$$

where q is the flow of water (cm^3/s)
   i is the hydraulic gradient causing the flow

$$i = \frac{H \text{ (head loss due to flow through soil)}}{L \text{ (length of path of flow through soil)}}$$

   A is the cross-sectional area of the flow path (cm^2)
and   *k* is the coefficient of permeability or average velocity of water through the soil (cm/s)

Since *i = H/L*, Darcy's law can also be stated as

$$q = k \frac{H}{L} A \qquad\qquad (1–14)$$

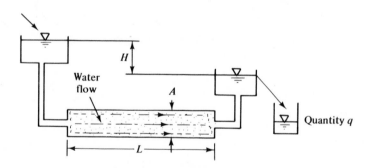

Water flow

H

A

Quantity *q*

L

FIG. 1–17.   Permeability of soils.

Example 1–19:   Referring to fig. 1–17, assume that

—   the coefficient of permeability $(k) = 0.75$ cm/s
—   the head loss as water flows through the soil $(H) = 6.50$ cm
—   the length of the flow path $(L) = 23.5$ cm
—   the cross-sectional area of the soil conduit $(A) = 5.8$ cm^2

The flow is

$$q = k \frac{H}{L} A$$

$$= 0.75 \ \frac{cm}{s} \times \frac{6.5 \ cm}{23.5 \ cm} \times 5.8 \ m^2$$

$$= 1.2 \ cm^3/s$$

Example 1–20: A canal and a river are parallel to each other at a distance of 200 m, with the canal being 7.5 m above the river. A 1.5-m layer of permeable sand (k = 0.085 cm/s) runs between them. Find the seepage from the canal to the river in m³/min per meter along the canal.

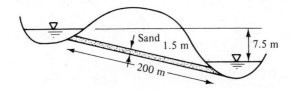

In this problems it is usually best to convert the given values to the units required in the answer before inserting them in the flow equation.

$$k = 0.085 \ \frac{cm}{s} \times \frac{1 \ m}{100 \ cm} \times \frac{60 \ s}{1 \ min} = 0.051 \ m/min$$

$H = 7.5 \ m$

$L = 200 \ m$

$A = 1.5 \ m \times 1 \ m = 1.5 \ m^2$ (cross-sectional area of conduit for a one-meter slice along the canal is 1.5 m high × 1 m wide)

$$q = k \frac{H}{L} A$$

$$= 0.051 \ \frac{m}{min} \times \frac{7.5 \ m}{200 \ m} \times 1.5 \ m^2 = 0.0029 \ m^3/min \text{ per meter of canal}$$

Example 1–21: A cofferdam (a temporary dam for construction purposes) is built across the mouth of a small bay. It is 135 ft. in length and is underlain by a permeable sand layer as shown. Water seeps through this sand (k = 0.064 cm/s).

Find seepage through the sand per linear foot of the dam, and total seepage for the whole dam, in cubic feet per minute.

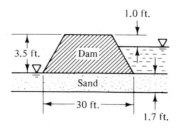

$$k = 0.064 \ \frac{cm}{s} \times \frac{1 \ ft}{30.48 \ cm} \times \frac{60 \ s}{1 \ min.} = 0.126 \ ft/min.$$

$L = 30$ ft, $H = 2.5$ ft, A: $1.7 \times 1 = 1.7 \ ft^2$ per foot of dam

$$\therefore q = 0.126 \ \frac{ft}{min} \times \frac{2.5 \ ft}{30 \ ft} \times 1.7 \ ft^2: \ 0.0178 \ ft^3/min \text{ per foot of dam}$$

Total seepage is $0.0178 \ ft^3/min/ft \times 135 \ ft = 2.4 \ ft^3/min$.

Note: In these problems the head loss and path length must be measured for the same flow path. The head loss between two points can be measured by finding the difference in elevation of head levels of the two points.

1—5.4   Typical values for the coefficient of permeability are listed in table 1—6. You can see that the permeability of soils varies tremendously, from very permeable gravels to impermeable (for all practical purposes) clays that may have a coefficient of permeability of 10^{-7} cm/s (about 0.1 ft/yr). The flow of water in pipes or conduits varies directly with the square of the size of the conduit. Gravels have very large grains, and therefore have large pores (spaces between grains) for water flow. Sands and silts have much smaller grains, and therefore much smaller pore spaces. Permeability is thus much less for sands and silts than for gravels—even though the total amount of pore space may be similar—since the individual pore spaces between grains are very small. In clays, the pore spaces are usually filled with attached water—which does not flow—leaving practically no effective pore space for water flow. Therefore clays, although they have high void ratios, are almost impermeable.

For any soil type, the coefficient of permeability depends on its density. The higher the density, the smaller the size of the pore spaces and, thus, the permeability.

Table 1–6
TYPICAL RANGES OF PERMEABILITY*

Soil Type	Relative Degree of Permeability	k, Coefficient of Permeability (cm/s)	Drainage Properties
Clean gravel	High	1 to 10	Good
Clean sand, sand and gravel mixtures	Medium	1 to 10^{-3}	Good
Fine sands, silts	Low	10^{-3} to 10^{-5}	Fair through poor
Sand-silt-clay mixtures, glacial tills	Very low	10^{-4} to 10^{-7}	Poor through practically impervious
Homogeneous clays	Very low to practically impermeable	Less than 10^{-7}	Practically impervious

Note: To convert cm/sec to ft/min, multiply cm/s by 2; i.e., 1 cm/s = 2 ft/min; also ft/day = cm/s × 3 × 10^3.

* David F. McCarthy, *Essentials of Soil Mechanics and Foundations,* 1977. Reprinted with permission of Reston Publishing Company, Inc., a Prentice-Hall Company, 11480 Sunset Hills Road, Reston VA 22090.

1–5.5 The coefficient of permeability for soil can be found as follows:

1. For clean, uniform sands, from Hazen's formula (an approximate value).

$$k = (D_{10})^2 \qquad\qquad (1-15)$$

where k is the coefficient of permeability (cm/s); D_{10} is the effective size (mm). This relationship was developed from the work of Hazen (1911) on sands.

Example 1–22: A clean, uniform sand has an effective size of 0.65 mm.

$$\therefore k = (.65)^2 = 0.42 \text{ cm/s}$$

2. For sands, from the *constant head permeability test.*
3. For fine sands and silts, from the *falling head permeability test.*
4. For clays, from the *consolidation test.*
5. For gravels and sands, from a *field test* using wells.

Only the two common laboratory tests (2 and 3) are described here.

In the *constant head permeability test,* water flows through a prepared sample of soil and the quantity is collected over a period of time. Since q, H, L, and A can be measured, the value of k can be calculated from Darcy's law (1–14), restated as $(k = qL/HA)$. A typical apparatus is shown in fig. 1–18.

Example 1–23: Using the apparatus shown in fig. 1–18, the following results were obtained:

$$H \text{ (head loss)} = 26.3 \text{ cm}$$

$$L \text{ (length of path)} = 17.1 \text{ cm}$$

$$A \text{ (cross-sectional area)} = 6.2 \text{ cm}^2$$

$$\text{flow} = 83.1 \text{ cm}^3 \text{ in 30 seconds}$$

$$\therefore q = 83.1/30 = 2.77 \text{ cm}^3/\text{s}$$

$$k = \frac{2.77 \times 17.1}{26.3 \times 6.2} = 0.29 \text{ cm/s}$$

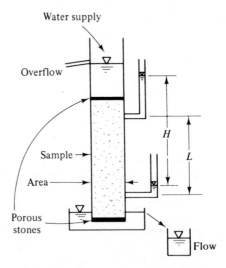

FIG. 1–18. Constant head permeameter.

The *falling head permeability test* is used for fine sands and silts, where the quantity of flow would be too small to measure properly by the constant head permeability test. An apparatus is shown in fig. 1–19. The sample is prepared and saturated. A head h_1 is applied to the sample. The flow control valve is opened and the drop in level to h_2 is measured in a certain time. The coefficient of permeability is calculated from

$$k = \frac{La}{TA} \ln \frac{h_1}{h_2} \qquad (1\text{--}16)$$

where a = area of standpipe
A = area of sample
T = time
L = length of sample
h_1, h_2 = initial and final heads

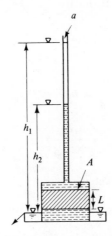

FIG. 1–19. Falling head permeameter.

Example 1–24: For the apparatus shown in fig. 1–19, a = 2.50 cm^2, A = 18.5 cm^2, L = 7.32 cm, T = 120 s, h_1 = 35.8 cm, and h_2 = 23.1 cm. Find k.

$$k = \frac{7.32 \text{ cm} \times 2.50 \text{ cm}^2}{120 \text{ s} \times 18.5 \text{ cm}^2} \ln \frac{35.8 \text{ cm}}{23.1 \text{ cm}}$$

$$= 0.0036 \text{ cm/s} \quad or \quad 3.6 \times 10^{-3} \text{ cm/s}$$

1–5.6 Capillary water is water that rises in tubes or pore spaces due to surface tension. To cite but one example, the *meniscus* that forms where water touches the side of a glass is due to surface tension.

The height that capillary water rises above the water table varies inversely with the diameter of the tube in which it rises. This principle can be demonstrated by the use of capillary tubes of various diameters.

Height of capillary water rise can be obtained by equating the force raising the water in a tube to the force of gravity on the volume of water raised.
Referring to fig. 1—20:

$$\text{Surface tension force} = S.T. \times \pi d \text{ (circumference of tube)}$$

$$\text{Force due to water} = \pi \frac{d^2}{4} \times h_c \times g \times \rho_W$$

where $S.T.$ = surface tension force
  d = diameter of capillary tube
  h_c = height of capillary rise
  g = acceleration due to gravity
  ρ_W = density of water

Equating these forces:

$$S.T. \times \pi d = \pi \frac{d^2}{4} \times h_c \times g \times \rho_W$$

or
$$h_c = \frac{4 \times S.T.}{d \times g \times \rho_W}$$

Using approximate values for water:

$$S.T. = 0.000735 \text{ N/cm} \quad or \quad 0.075 \text{ g/cm}$$

$$g\,\rho_W = 0.0098 \text{ N/cm}^3 \quad or \quad 1 \text{ g/cm}^3$$

$$h_c = \frac{4 \times 0.000735 \text{ N/cm}}{d \times 0.0098 \text{ N/cm}^3} \quad or \quad \frac{4 \times 0.075 \text{ g/cm}}{d \times 1 \text{ g/cm}^3}$$

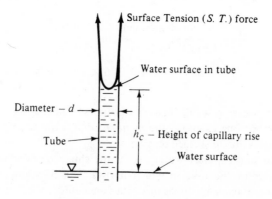

FIG. 1—20. Capillary rise.

If d is expressed in centimeters:

$$h_c \text{ (cm)} = \frac{0.3}{d(\text{cm})} \qquad (1-17)$$

Example 1–25: Find the height of capillary rise in tubes with diameters of (a) 1 cm, (b) 1 mm, and (c) 0.005 mm.

(a) $h_c = 0.3/1$ cm = 0.3 cm
(b) $h_c = 0.3/0.1$ cm = 3 cm
(c) $h_c = 0.3/0.0005$ cm = 600 cm

1–5.7 The pore spaces in soils are similar to tubes. Water rises above the groundwater table in these pores. Pore sizes vary greatly in a soil, and are difficult to estimate reliably. A value of 20% of the effective size is often used to approximate the pore size. Therefore, for a soil with an effective size at about the No. 200 sieve size (0.075 mm or 0.0075 cm), the average pore size might be 0.0075 × 0.20 = 0.0015 cm, and the height of capillary rise might be 0.3/0.0015 = 200 cm (about 7 ft). Typical values for height of capillary rise are:

Sands	0–1 m (0–3 ft.)
Silts	1–10 m (3–33 ft.)
Clays	over 10 m (33 ft.)

This surface tension in soil water, which causes water to rise in capillary spaces, has three important effects.

1. Soil is saturated for a distance above the groundwater table, as shown in fig. 1–15. Saturation in soils cannot be used as an indication of the location of the groundwater table.
2. Apparent cohesion in silts and sands is due to the surface tension forces in the moisture film surrounding the soil grains. Silt grains usually stick together when found in deposits. However, this apparent cohesion disappears when the soil is dried or fully saturated. Sands tend to bulk when piled in a moist condition, due to the surface tension forces holding the grains and resisting their movement to a denser configuration. Figure 1–21 illustrates this action.
3. Frost heaving is a major problem in Canada and the northern United States, since it causes many pavement failures. Steps in the growth of a frost heave (as illustrated in fig. 1–22) are as follows:
    a. As a freezing front descends, water in large pores freezes, forming an ice crystal.

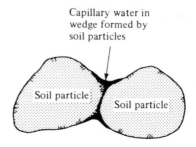

FIG. 1–21. Apparent cohesion between soil grains due to surface tension in the moisture film. (David F. McCarthy, *Essentials of Soil Mechanics and Foundations*, 1977. Reprinted with permission of Reston Publishing Company, Inc., A Prentice-Hall Company, 11480 Sunset Hills Rd., Reston, VA 22090.)

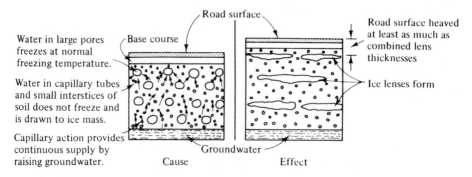

FIG. 1–22. Mechanics of frost heave. (Reprinted with permission from Clarkson H. Oglesby, *Highway Engineering*, 3rd ed. New York: Wiley, 1975.)

b. Water in adjacent, smaller capillary pore spaces does not freeze immediately; its freezing temperature is slightly lower due to the very small size of the water particles.

c. This adjacent water is drawn to the ice crystal already formed; it then freezes, enlarging the size of the crystal.

d. Capillary water moves up the capillary tube to replace the water that has joined the ice crystal; this capillary water also moves to the ice crystal and freezes.

e. This continual flow of water from the groundwater table to the ice crystal results in the growth of an ice lens, which may grow up to 15 cm (6 in.) in thickness.

    f. As the freezing front penetrates deeper into the capillary zone, more ice lenses may form.

    g. The surface soil heaves by an amount equal to the total thickness of the ice lenses formed.

(Frost heaving is discussed further in chapter 5.)

1–6 SOIL STRENGTH AND SETTLEMENT

The two main types of failure that occur in soils are (1) failures due to *shear,* that is, where grains slide with respect to other grains, and (2) *settlement* failures, that is, where a layer of soil is compressed and becomes thinner under loading. These failures are illustrated in fig. 1–23.

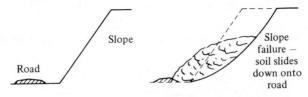

(a) Shear Failure

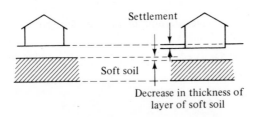

(b) Settlement Failure (amount of settlement equals decrease in thickness of layer of soft soil)

FIG. 1–23. **Types of failure in soils.**

1–6.1 A full analysis of *shear strength* in soils is beyond the scope of this text. It would involve discussion of both stresses developed in pore water and combinations of shear and normal stresses.

    However, for the purpose of conducting routine soils evaluation and understanding the relevance of the field tests conducted on soils, the following introduction to shear strength will suffice.

Forces acting on a plane are normal forces (N), which act perpendicular to the plane, and shear forces (S), which act parallel to the plane. These are illustrated in fig. 1—24 showing forces acting on a block sliding on a surface. Normal and shear forces are usually measured in kN or pounds.

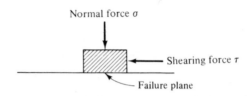

Normal force σ

Shearing force τ

Failure plane

FIG. 1—24.   Shear strength due to sliding friction.

Stress or force per unit area is found by dividing the total force by the area on which it acts. Normal stress (σ) and shear stress (τ) have units of kN/m^2 (kPa) or lb/ft^2.

Shear strength is shear stress resisting failure along a plane, as illustrated in fig. 1—25.

Shear force

Resisting force

FIG. 1—25.   Shear strength due to sliding friction.

Shear strength in clays is due to cohesion between the grains, holding them together. For clays

$$\tau = c \qquad\qquad (1-18)$$

where τ is shearing resistance (kPa or lb/ft^2)
c is cohesion of soil (kPa or lb/ft^2)

In granular soils, shear strength results from friction between the grains along the shearing plane. This is similar to sliding friction produced as a block slides across a table, as shown in fig. 1—24. The shearing pressure required to cause sliding (τ) varies with the mass of block, or the normal stress on the plane of failure (σ). For example:

For σ = 10 kPa, τ might be 6 kPa

For σ = 20 kPa, τ would be 12 kPa

For σ = 30 kPa, τ would be 18 kPa

By plotting these stresses on stress axis as shown in fig. 1–26, a failure line is obtained giving shear stress at failure (or shear strength) corresponding to any value of normal stress. The angle that this line makes with the horizontal is the *angle of internal friction*, ϕ.

$$\text{Tan } \phi = \tau/\sigma$$

or
$$\tau = \sigma \tan \phi$$

where τ is shearing resistance kPa (lb/ft^2) (1–19)
    σ is normal load on plane of failure kPa (lb/ft^2)
    ϕ is angle of internal friction

(For the values plotted in fig. 1–26, $\phi = 31°$.)

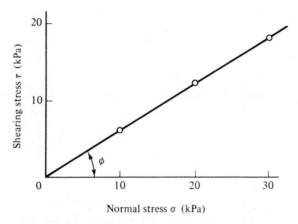

FIG. 1–26. Plot of shear and normal stresses.

To summarize, shear strength in most clays is due to cohesion, and $\tau = c$. Shear strength in granular soils is due to friction and $\tau = \sigma \tan \phi$. The cohesion, c, and the friction angle, ϕ, of the soil are measured in various shear strength tests.

Mixed soils and partially saturated or hard clays may have shear strength developed by both cohesion and friction. In this case,

$$\tau = c + \sigma \tan \phi$$ (1–20)

1–6.2 Shear strength in soils is measured by:

— unconfined compression test (clays only)
— direct shear test

- triaxial compression test
- vane shear test (clays only)
- field tests, described in chapter 2

In the *unconfined compression test,* a cyclindrical sample is prepared, measured, and compressed in a compression apparatus. The strain (change in length) and the load carried by the sample at failure are recorded. The strain at failure, ϵ_f, expressed as a ratio of the original length, is used to correct the cross-sectional areas as follows:

$$A_f = \frac{A_o}{1 - \epsilon_f}, \qquad\qquad (1-21)$$

where A_o is the original area and A_f is the corrected area at failure. The unconfined compressive strength (q_u) is the load per unit area,

$$q_u = \frac{\text{Max load}}{A_f} \qquad\qquad (1-22)$$

The shear strength (or cohesion) is one-half of the unconfined compressive strength.

$$\tau = \frac{q_u}{2} \qquad\qquad (1-23)$$

Example 1–26: A sample of soil 4.0 cm in diameter and 7.5 cm in length, is tested in an unconfined compression test apparatus. Failure occured at a strain of 0.57 cm. Maximum load was 192 N. Find shear strength.

$$A_o = 12.6 \text{ cm}^2 = 0.00126 \text{ m}^2$$
$$\epsilon_f = 0.57 \text{ cm}/7.5 \text{ cm} = 7.6\%$$
$$A_f = 0.00126 \text{ m}^2/(1-0.076) = 0.00136 \text{ m}^2$$
$$q_u = 192 \text{ N}/0.00136 \text{ m}^2 = 141 \text{ kPa}$$
$$\tau = 141 \text{ kPa}/2 = 70 \text{ kPa}$$

A definite shear plane usually develops in the sample at an angle of about 55° to 60° with the horizontal, as shown in the figure on p. 50. In soft, saturated clays, the sample may bulge during the test without the development of a failure plane. The load on the sample keeps increasing as the sample is compressed. In this case failure is assumed to occur when the strain reaches 20%.

Example 1–27: An unconfined compression test is conducted on a soft clay sample, with a diameter of 1.40 in and a length of 2.90 in. The sample bulged during the test without reaching a maximum value. At a strain of 20% (0.58 in), the load was 17.05 lb. Find the shear strength.

$$A_o = 1.54 \text{ in.}^2$$
$$A_f = 1.54 \text{ in}^2/(1-0.20) = 1.92 \text{ in.}^2$$
$$q_u = 17.05/\text{lb}/1.92 \text{ in.}^2 = 8.88 \text{ lb/in}^2$$
$$\tau = 8.88 \text{ lb/in}^2/2 = 4.44 \text{ lb/in}^2 = 640 \text{ lb/ft}^2$$

The *direct shear test* uses a shear box that is divided in half horizontally. The soil sample is placed in the box, a normal load (N) is applied on the top surface and the one half of the box is forced to slide over the other by a shear force (S). The maximum value of the shear force is measured. Stresses at failure, σ and τ, are then calculated. For cohesionless soils, the friction angle can be found directly using equation (1–19). For soft clays, the shear strength is the shear stress recorded in the test, and does not vary greatly with changes in the normal load applied. In the case of mixed soils two tests must be conducted so that c and ϕ can be found by plotting the test results, as shown in example 1–29.

Example 1–28: In a direct shear test on a sand soil with a normal load of 50 lb., maximum resistance during shearing of the sample was measured as 31.7 lb. The shear box is 2.0 in. square. Find shear strength.

$$\sigma = 50 \text{ lb}/4.0 \text{ in}^2 = 12.5 \text{ lb/in}^2$$
$$\tau = 31.7 \text{ lb}/4.0 \text{ in}^2 = 7.92 \text{ lb/in}^2$$
$$\phi = \arctan (7.92/12.5) = 32°$$

Shear strength, $\tau = \sigma \tan 32°$

Example 1–29: Two tests are conducted on a mixed soil in a direct shear test, with normal loads of 100 and 200 N. In test 1, maximum shear resistance was 92 N, in test 2, 112 N. The shear box was 5.00 cm in diameter. Find shear strength.

$$\text{Area of shear box} = .00196 \text{ m}^2$$

Test 1, $\sigma = 100/.00196 = 51$ kPa,
$\tau = 92/.00196 = 47$ kPa

Test 2, $\sigma = 200/.00196 = 102$ kPa,
$\tau = 112/.00196 = 57$ kPa

By plotting these values, the cohesion of the soil is found to be 37 kPa and the friction angle 11°. The shear strength is:

$$\tau = (37 + \sigma \tan 11°) \text{ kPa}$$

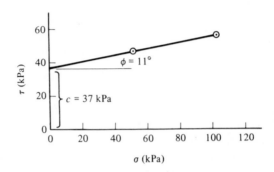

The *triaxial compression test* is a confined compression test. The sample is enclosed in a rubber membrane and placed in a triaxial cell. Water is introduced into the cell, and a pressure, σ_3 pressure, is applied to the water to confine the sample in the same manner that it is confined in the field by soil all around it. The sample is sheared by a vertical force, σ_1.

The triaxial test can be conducted with or without drainage occurring and with or without measurements being taken of stress in the soil water. Clays are often tested in a quick shear test in the triaxial apparatus, without drainage or water pressure measurements. Cell pressure, σ_3, is applied to the sample. Then an additional axial load is applied until failure. The additional axial load at failure, $(\sigma_1 - \sigma_3)$, divided by the corrected area of the sample (found as in the unconfined compression test), is the compressive strength in quick shear, (q_q). Shear strength is 50% of this value.

Example 1–30: A clay sample 3.5 cm in diameter and 6.10 cm in length is prepared and tested in quick shear in a triaxial compression apparatus. A cell pressure of 30 kPa, representing the confining stress on the soil in the field, is applied. Then an additional axial load is applied until failure. At failure this was 73.5 N and the axial strain was 0.49 cm. Find shear strength.

$$A_o = 9.62 \text{ cm}^2$$

$$\epsilon_f = 0.49 \text{ cm}/6.10 \text{ cm} = 8.0\%$$

$$A_f = 9.62 \text{ cm}^2/(1 - .080) = 10.5 \text{ cm}^2$$

$$q_q = 73.6 \text{ N}/.00105 \text{ m}^2 = 70 \text{ kPa}$$

$$\tau = 70 \text{ kPa}/2 = 35 \text{ kPa}$$

In the *vane shear test,* a vane with four blades is inserted into the sample. The torque required to rotate the vane is measured, and the cohesion of the soil is calculated from the results.

Schematic diagrams of three of these tests are shown in fig. 1–27. A photograph of an unconfined compression test is shown in fig. 1–28.

1–6.3 The amount that a building or structure might settle is governed by the *compressibility* of the soil. Compressibility involves the rearrangement of the soil grains to a denser, thinner layer, usually involving the squeezing out of water when a load is placed on the soil (see fig. 1–29).

Settlement is a serious problem in some types of clay:

1. Clays may have a loose structure and a high void and moisture content, and can therefore be compressed considerably.
2. Due to the extremely slow movement of water in clays, the time required for settlement to take place may be years.

In granular soils, the grains are usually in close contact. In any event, any settlement usually takes place as the load is being applied and does not lead to long-term settlement problems.

The amount and rate of compressibility—or *consolidation,* as it is usually called in saturated clays—are calculated from results of the *consolidation test.* In this test a sample of soil is placed in a cell, and a load is placed on it. Thickness of the sample is measured at the start of the loading, and then periodically for 24 hours. Then the load is doubled, and again thickness is recorded. The test proceeds for four or five loadings. Using these results and knowing the magnitude of the proposed load and the depth and thickness of the clay layer, you can calculate the amount and rate of settlement.

1–7 TEST PROCEDURES

Procedures for the following tests are briefly outlined. (For detailed test procedures, refer to the standards listed in the Appendix.)

1–7.1 Relative Density (Specific Gravity) of Soils
1–7.2 Mass-Volume Measurements
1–7.3 Grain Size by Sieve Analysis
1–7.4 Grain Size by Hydrometer Analysis
1–7.5 Atterberg Limits Test
1–7.6 Constant Head Permeability Test
1–7.7 Unconfined Compression Test
1–7.8 Direct Shear Test

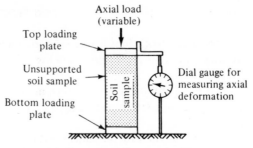

(a) Unconfined Compression Test

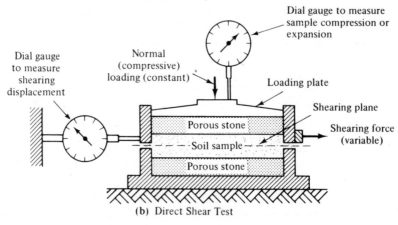

(b) Direct Shear Test

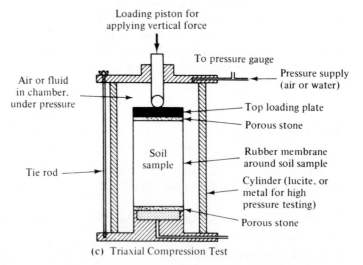

(c) Triaxial Compression Test

FIG. 1–27. Shear strength tests. (David F. McCarthy, *Essentials of Soil Mechanics and Foundations*, 1977. Reprinted with permission of Reston Publishing Company, Inc., a Prentice-Hall Company, 11480 Sunset Hills Rd., Reston, Va. 22090.)

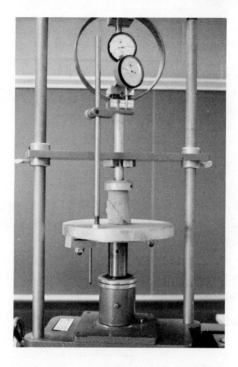

FIG. 1−28. Unconfined compression test.

1−7.1 *Relative Density (Specific Gravity) of Soils*

Purpose: To measure relative density of a soil.

Theory: Relative density (specific gravity) of a soil is (1) the ratio of the mass of the soil to the mass of an equal volume of distilled water or (2) the ratio between the density of soil particles and the density of water. The density of

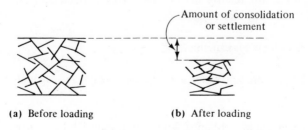

(a) Before loading   (b) After loading

FIG. 1−29. Clay structure before and after consolidation.

water is 1.000 g/cm^3 at $4°C$, the temperature at which water is in its densest state. The relative density of a soil is usually reported relative to water at $20°C$.

Apparatus: pycnometer (100 ml)
                balance (accurate to 0.01 g)
                vacuum pump
                oven

Procedure:

1. Oven-dry the soil.
2. Place about 25 g in the pycnometer and find the mass to within 0.01 g.
3. Add water until the pycnometer is about three-fourths filled.
4. Apply a partial vacuum to the sample to remove any air.
5. Fill with water to the calibration mark on the pycnometer. Obtain the mass of the pycnometer.
6. Record the temperature of the water in the pycnometer.

Results and calculations:

Pycnometer No.	_____	
Mass of soil plus pycnometer	_____	g
Mass of pycnometer	_____	g
Mass of soil (dry)	_____	g (M_o)
Mass of pycnometer, soil, and water	_____	g (M_b)
Temperature	_____	$°C$ (Tx)
Mass of pycnometer and water at Tx	_____	g (M_a)
Relative density at Tx (specific gravity)	_____	$RD = \dfrac{M_o}{M_o + (M_a - M_b)}$

Note: The value for relative density should be corrected to $20°C$ if the test temperature varies significantly.

1–7.2   Mass-Volume Measurements

Purpose: To measure soil, water, and air constituents, and calculate mass-volume relationships.

Theory: A soil sample obtained with a thin-walled tube in a relatively undisturbed state can be measured to obtain the volume and mass of soil, water, and air. From these, the density, dry density, water content, void ratio, porosity, and degree of saturation can be calculated. (The relative density of the soil must be known.)

Apparatus: oven
            balance (accurate to 0.01 g)

Procedure:

1. Extract a sample 20-50 mm (1-2 in) in length from a thin-walled soil sampling tube.
2. Measure the diameter and the length.
3. Place in a beaker and find the mass (to within 0.01 g).
4. Place in an oven at 110°C (±5°) for 20-30 hours to evaporate water.
5. Find the mass of the dry soil and the beaker.

Results:

Sample diameter _____ cm
length _____ cm
volume _____ cm^3
Mass: wet sample plus beaker _____ g
dry sample plus beaker _____ g
water _____ g
beaker _____ g
dry sample _____ g

Calculations: (RD =   )

$$V = \underline{\hspace{3cm}} \text{ cm}^3$$

$$M_D = \underline{\hspace{3cm}} \text{ g}$$

$$M_W = \underline{\hspace{3cm}} \text{ g}$$

$$V_D = \frac{M_D}{RD \times 1 \text{ g/cm}^3} = \underline{\hspace{2cm}} = \underline{\hspace{2cm}} \text{ cm}^3$$

$$V_W = \frac{M_W}{1 \text{ g/cm}^3} = \frac{\underline{\hspace{2cm}}}{1} = \underline{\hspace{2cm}} \text{ cm}^3$$

$$V_A = V - (V_D + V_W) = \underline{\hspace{2cm}} - \underline{\hspace{2cm}} = \underline{\hspace{2cm}} \text{ cm}^3$$

Therefore:

$$\rho = \underline{\hspace{3cm}} \qquad e = \underline{\hspace{3cm}}$$
$$\rho_D = \underline{\hspace{3cm}} \qquad n = \underline{\hspace{3cm}}$$
$$w = \underline{\hspace{3cm}} \qquad S = \underline{\hspace{3cm}}$$

1–7.3 Grain Size by Sieve Analysis

Purpose: To obtain grain size distribution of a soil by a sieve analysis.

Theory: A sample of soil is dried so that (1) weights obtained are of soil parti-cles only, and (2) grains are not bound together by surface tension in water film. The sample is placed in a nest of sieves that are arranged in order of size of opening (large to small from top to bottom), then shaken by rotary and up-and-down motion until all grains have passed through all sieves possible according to size. The size of the sample must be (1) large enough to be representative of the soil being tested, but (2) small enough so that no sieve is covered by a thick layer of soil grains that would prevent some grains from going through that sieve. If sieving is completed dry, the percentage passing 75 μm (No. 200) may not be accurate. To find the correct percentage passing 75 μm (No. 200), soil particles are washed over the sieve to wash fine particles through.

Apparatus: sieves
sieve shaker
balances

Procedure:

1. Oven-dry the sample.
2. Measure the mass of the dried sample.
3. Place in the nest of sieves and shake for five minutes.
4. Measure mass retained on each sieve.

Results:

Original mass of soil_____ g

Calculations:

1. Calculate the percentages of gravel, sand, and fines (clay and silt). (Gravel is larger than 4.75 mm (No. 4); sand, 4.75 mm to 75 μm (No. 200); fines, smaller than 75 μm.)
2. Calculate effective size (D_{10}) and uniformity coefficient ($C_u = D_{60}/D_{10}$).

Results of sieving:

Sieve No.	Mass retained	% Retained	Cumulative % passing
Totals			

1-7.4   Grain Size by Hydrometer Analysis

Purpose: To determine distribution of particle sizes in a soil sample composed of fine-grained soil sizes.

Theory: A known mass of soil is broken up and dispersed uniformly in a cylinder of water. Readings are taken with a hydrometer to determine the density of the soil-water mixture. Using Stokes' law, it is possible to calculate the diameter D of a soil particle such that all coarser particles have already settled a distance L (surface to center of bulb) in time T, while all finer particles that originally were at the surface are still in suspension. Using the hydrometer reading for the density of the solution at the bulb, it is possible to find the percentage of the original sample that is still in suspension. The data give the percentage of particles finer than various particle diameters, and allow the plotting of a grain size distribution curve.

Hydrometers are calibrated to read "0" in distilled water at standard temperature at the surface of the water. Since (1) pure water is not used, (2) temperature varies, and (3) the reading must be taken at the top of the meniscus, a correction factor must be applied to each reading. A graph is usually available in the laboratory to give the correction factor, which varies with the temperature.

In this test a solution is added to neutralize the bonds between grains, and the sample is mixed in a "milk shake" apparatus to break up clumps of grains. Then the sample is placed in a jar and mixed to ensure that the grains are distributed uniformly in the jar. The jar is then set down, and the grains are allowed to settle.

Apparatus: hydrometer
             hydrometer jar
             stirring apparatus
             sieves–2.00 mm (No. 10), 75 μm (No. 200)
             balance (accurate to 0.01 g)
             thermometer
             dispersing agent (a solution of 40 g sodium hexametaphosphate in
               1 L solution)

Procedure:

1. Oven-dry the sample, break it down in a mortar, and pass it through a
   2.00 mm (No. 10) sieve.
2. Place 50–60 g in beaker, and obtain the mass to within 0.01 g.
3. Add 125 ml of dispersing agent. Stir, rinsing any soil grains off the spatula
   with a wash bottle. Allow the sample to soak (which neutralizes the bonds
   between grains).
4. Transfer to a dispersion cup while rinsing all the soil in. Fill the cup about
   halfway, and stir for approximately one minute.
5. Transfer the soil to the hydrometer jar, rinsing the cup out with a wash
   bottle. Fill the jar to the 1000 ml (1 L) mark with water.
6. With one's hand over the end, turn the jar vigorously for one minute.
7. Set the jar down immediately. Note and record the time immediately as the
   "start" of the test.
8. Take hydrometer and temperature readings. Suggested times are 1, 2, 5, 15,
   30, 60, 250, and 1440 minutes after the start. Take readings at one and two
   minutes to the nearest second; others, by clock. Readings do not have to be
   exactly at the above intervals, but times to the nearest minute must be
   recorded.
9. At the end of the test, pour the sample out over the 75 μm (No. 200) sieve.
   Wash fines through sieve. Transfer retained material to the beaker, dry, and
   obtain mass.

Results:

                    Original mass of sample_____ g (S)
                        Hydrometer test
             Washing on 75 μm sieve
                 mass retained _____ g
                 mass passing _____ g
                 percentage passing _____ %
             Type of hydrometer 151H _____ or 152H _____
             Relative density of soil_____(RD)

Calculations:

1. Percentage passing 75 μm (No. 200) = mass passing/S.
2. Complete the table of results.

Time: Start of Test [   ]

Clock time	Elapsed time T–min	Temp °C	Hyd. reading R'	Hyd. corr. C	Correct reading R	L	K	Particle dia. D mm	% Smaller P

(a) Find the hydrometer correction C for each reading and, adding or subtracting as indicated, find the correct reading R (from chart available in lab).
(b) Find the effective depth L from table 1–7A or 1–7B.
(c) Find the values for K from fig. 1–30.
(d) Calculate particle diameter $(D = K\sqrt{L/T})$.
(e) Find the relative density correction factor a from table 1–8 (type 152H only).
(f) Calculate the percentage smaller P:

$$P = \frac{Ra}{S} \times 100 \quad \text{(type 152H)}$$

or

$$P = \frac{100,000}{S} \times \frac{RD}{RD - 1}\,(R - 1) \quad \text{(type 151H)}$$

Conclusion: Draw the grain size distribution curve using results from the hydrometer test and the percentage passing the 75 μm (No. 200) sieve.

Table 1-7A
VALUES FOR EFFECTIVE DEPTH, *L*, FOR TYPE 151H HYDROMETER *

Actual Hydrometer Reading	L (cm)	Actual Hydrometer Reading	L (cm)
1.000	16.3	1.020	11.0
1.001	16.0	1.021	10.7
1.002	15.8	1.022	10.5
1.003	15.5	1.023	10.2
1.004	15.2	1.024	10.0
1.005	15.0	1.025	9.7
1.006	14.7	1.026	9.4
1.007	14.4	1.027	9.2
1.008	14.2	1.028	8.9
1.009	13.9	1.029	8.6
1.010	13.7	1.030	8.4
1.011	13.4	1.031	8.1
1.012	13.1	1.032	7.8
1.013	12.9	1.033	7.6
1.014	12.6	1.034	7.3
1.015	12.3	1.035	7.0
1.016	12.1	1.036	6.8
1.017	11.8	1.037	6.5
1.018	11.5	1.038	6.2
1.019	11.3		

* Reprinted from ASTM Standard D422-63 by permission of the American Society for Testing and Materials, 1916 Race Street, Philadelphia PA 19103, Copyright.

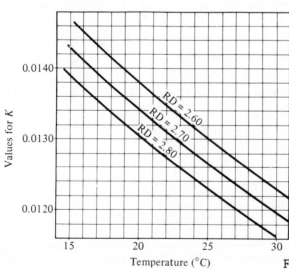

FIG. 1-30. Values for K.

Table 1−7B

VALUES FOR EFFECTIVE DEPTH, L, FOR TYPE 152H HYDROMETER*

Actual Hydrometer Reading	L (cm)	Actual Hydrometer Reading	L (cm)
0	16.3	31	11.2
1	16.1	32	11.1
2	16.0	33	10.9
3	15.8	34	10.7
4	15.6	35	10.6
5	15.5		
6	15.3	36	10.4
7	15.2	37	10.2
8	15.0	38	10.1
9	14.8	39	9.9
10	14.7	40	9.7
11	14.5	41	9.6
12	14.3	42	9.4
13	14.2	43	9.2
14	14.0	44	9.1
15	13.8	45	8.9
16	13.7	46	8.8
17	13.5	47	8.6
18	13.3	48	8.4
19	13.2	49	8.3
20	13.0	50	8.1
21	12.9	51	7.9
22	12.7	52	7.8
23	12.5	53	7.6
24	12.4	54	7.4
25	12.2	55	7.3
26	12.0	56	7.1
27	11.9	57	7.0
28	11.7	58	6.8
29	11.5	59	6.6
30	11.4	60	6.5

* Reprinted from ASTM Standard D422−63 by permission of the American Society for Testing and Materials, 1916 Race Street, Philadelphia PA 19103, Copyright.

Table 1-8

VALUES FOR RELATIVE DENSITY CORRECTION FACTOR, a, FOR TYPE 152H HYDROMETER*

RD	a
2.80	0.97
2.75	0.98
2.70	0.99
2.65	1.00
2.60	1.01
2.55	1.02

* Reprinted from ASTM Standard D422-63 by permission of the American Society for Testing and Materials, 1916 Race Street, Philadelphia PA 19103, Copyright.

1-7.5 Atterberg Limits Test

Purpose: To obtain values for liquid limit, plastic limit, and index of plasticity.

Theory: The amount of water a soil sample can absorb before changing from a solid to a plastic state and then to a liquid state is a very important indication of (1) whether the soil is mainly silt or clay, and (2) if it is clay, the characteristics of the particular clay minerals in the sample.

The plastic limit is the moisture content at which the sample changes from a solid material to one that is plastic. It is defined as the moisture content at which it is possible to roll the soil to a thread about 3 mm (1/8 in) in diameter without having the soil crumble.

The liquid limit is the moisture content at which the sample changes from a plastic state to a liquid state. It is defined as the moisture content at which the soil will flow together at 25 drops of a cup, after having been separated with a groove.

Apparatus: liquid limit device (check drop to be 1 cm)
plastic limit plate
balance (accurate to 0.01 g)
evaporating dish and petri dishes

Procedure:

1. Air-dry the sample, break it down in a mortar, and sieve through a 425 μm (No. 40) sieve.
2. Place 125-150 g of the sample in an evaporating dish.

3. Add small increments of water, mixing thoroughly each time with a spatula by stirring, kneading, and chopping actions. Add water until the sample is between plastic and liquid limits.

4. Take part of the sample, roll it into a ball, and roll it on the glazed surface of a plate. If it is too wet, roll it on the rough surface of the plate to remove excess water. Roll on a glazed surface to a thread of about 3 mm (1/8 in) diameter, break the thread into pieces, squeeze pieces together, and re-roll. Roll and re-roll the thread until it crumbles under the pressure required to roll it to a 3 mm thread. It is then at its plastic limit. Put the thread in a petri dish and obtain the mass (to within 0.01 g). Record the container number and mass of the wet sample plus container.

5. Place a part of the sample in the liquid limit cup, squeeze it down, and smooth it out with a few strokes of the spatula. The sample should be about 1 cm deep at the center.

6. Divide the sample in the cup along the center with a grooving tool.

7. Lift and drop the cup by its handle, counting the drops until the sample comes together over a 13 mm (1/2 in) length along the base of the groove.

8. If the soil comes together after 5–50 drops, take a sample of the soil through the center, place it in a petri dish, and obtain the mass (to within 0.01 g). Record the number of drops, dish number, and mass of the wet sample and container. (If the sample requires more than 50 drops to come together, it is too dry; if it requires fewer than five, it is too wet.)

9. Transfer the soil to an evaporating dish. Wipe out the cup. Add a little water (or allow to dry), mix thoroughly, and test again. Repeat three or more times until there are at least three samples taken—one at 30–50 drops, one at 5–20 drops, and one in between.

10. Dry the samples. Obtain the mass of the dried samples, and calculate all water contents.

Results:

Plastic limit
  trial no.   _____   _____   _____
  w _____   _____   _____
Liquid limit
  trial no.   _____   _____   _____   _____   _____
  number of drops _____   _____   _____   _____   _____
  w _____   _____   _____   _____   _____

Calculations:

1. The plastic limit is the average of values obtained in the plastic limit tests.

2. Plot the results of the liquid limit test on a graph. Draw a straight line through the test points. The liquid limit is the water content where the test line intersects the 25-drop line.

3. Index of plasticity $(I_p) = w_L - w_P$.

Liquid limit flow graph

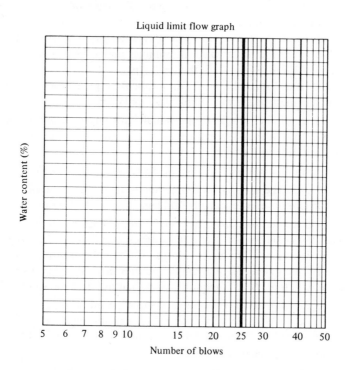

Number of blows

1-7.6 Constant Head Permeability Test

Purpose: To measure permeability of a sand soil.

Theory: Water flows through a soil sample in a tube. Head on the water is kept constant. The water that flows through in a known time is collected. Using Darcy's law, the coefficient of permeability can be calculated.

Apparatus: permeameter

Procedure:

1. Place porous stone in the bottom of the permeameter.
2. Obtain the mass of the sand and container.

3. Add sand to the permeameter, compacting it to the desired density. Place porous stone on the surface. Tap the stone to level the surface and eliminate air pockets.

4. Obtain the mass of the remaining sand.

5. Attach a water supply and allow it to flow through the sample until the rate of flow becomes uniform.

6. Measure the length L, head H, and tube's diameter.

7. With a beaker, catch the flow for a measured time.

8. Record the temperature of the water. (It should be approximately 20°C, or else a correction factor should be applied for the coefficient of permeability, k.)

Results:

1. Mass of sand used_____ g

2. $H =$ _____ cm $L =$ _____ cm Diameter =_____ cm

3. Total flow_____ cm^3 in_____ s

4. Temperature_____ °C

Calculations:

1. Area $A =$_____ cm^2

2. Flow $q =$_____ cm^3/s

3. Density of sand $= \dfrac{\text{mass}}{\text{volume}} = \dfrac{\text{mass}}{L \times A} =$_____ $=$_____ g/cm^3

4. Coefficient of permeability

$$k = \frac{qL}{HA}$$

$k =$ _____ $=$_____ cm/s

1–7.7 Unconfined Compression Test

Purpose: To find cohesion or shear strength of a cohesive soil.

Theory: This test finds the shear strength or cohesion of a cohesive (clay) soil, which has little or no friction strength. A sample from a thin-walled tube is trimmed to remove disturbed outer edges, and its ends are squared. It is tested in compression, without being confined on the sides, until failure. Readings are taken for the total load on the sample, as well as the axial strain or decrease in the length of the sample during the test. The test is over (1) when the sample fails, or (2) when the axial strain in soft soils, which only bulge during failure, is 20% of the original sample length. The load at failure, divided by the corrected area, gives the unconfined compressive strength (q_u). The shear strength or cohesion equals one-half the unconfined compressive strength.

Apparatus: compression machine
                  soil trimmer
                  mitre box
                  scales

Procedure:

1. Extract a sample about 80 mm (3 in) in length from a tube.
2. Trim the sample and measure its diameter.
3. Cut the ends in a mitre box so that the test sample's length is about twice its diameter.
4. Place the sample in an unconfined apparatus.
5. Measure the length (L_o).
6. Bring the loading head to the surface of the sample.
7. Set the load and strain gauges at zero.
8. Start compression.
9. Read the strain and load dials at the indicated times.
10. Continue the test until the specimen fails or strain is over 20%.
11. Sketch the sample at failure.

Results: Record load and strain at various times during the test.

Calculations:

1. Calculate the percentage of strain (total strain/L_o) for each reading.
2. Plot the percentage of strain vs. load. Draw the curve.
3. Locate the failure point at the top of the curve or, if it does not peak, at a strain of 20%.
4. At the failure point, read maximum load_____ (M)
               and strain_____ % (ϵ_f)
5. Calculate cross-sectional area   _____ (A_o).
6. Calculate area at failure $A_o/[1 - (\epsilon_f/100)]$ _____ (A_f).

7. Calculate unconfied compressive strength: M/A_f _____ (q_u).
8. Calculate shear strength: $q_u/2$ _____ τ.

1–7.8   Direct Shear Test

Purpose: To measure the shear strength of a soil under various normal loads, and to establish values for cohesion (c) and the angle of internal friction (ϕ).

Theory: In the direct shear test, a soil sample is caused to fail along a horizontal plane through the sample. The normal load of the plane of failure is known, and the maximum shear force required to cause shear or sliding of one part of the sample over another is recorded. If a series of tests is conducted, and the unit shear stresses (τ) plotted against the unit normal stresses (σ), values for cohesion and the angle of internal friction can be obtained. For cohesion less sands, $\phi = \arctan \tau/\sigma$ and for soft clays, the shear stress, τ, is the cohesion or shear strength.

Apparatus: direct shear box, consisting of upper and lower sections
direct shear machine

Procedure:

1. Place an undisturbed or carefully prepared sample in the shear box, with porous stones on the top and bottom (for drained tests only).
2. Apply normal load.
3. Fill the reservoir with water.
4. Unlock the pins holding the upper and lower parts of the shear box together, and separate the parts slightly.
5. Apply shear force and shear the sample slowly until the maximum shear force is reached.

Results:

Area of failure plane	_____	(A)
Normal load	_____	(N)
Maximum shear force	_____	(S)

Calculations:

Normal stress (N/A)	_____	σ
Shear stress (S/A)	_____	τ

1–8 PROBLEMS

Note: Problems 1–1 to 1–39 involve no units or only metric units commonly used in soils laboratories. (Students working in traditional units should substitute "unit weight" and "specific gravity" for "density" and "relative density," respectively.) Problems 1–40 to 1–62 are in *SI* units and are repeated in traditional units in problems from 1–63 to the end.

1–1. List the main soil types. Describe how they differ with regard to (a) grain size, (b) grain shape, and (c) forces between grains.

1–2. Surface properties of the grain are more important in clays than in silts. Why?

1–3. How do grain structures differ between deposits of clay and deposits of sands? Why?

1–4. Why do clays absorb a relatively large quantity of water?

1–5. List five methods of testing for silt and clay in the field. Describe how you would tell the difference between silt and clay.

1–6. (a) The mass of a quantity of water is 32.7 g. What is its volume in cm^3?

(b) The volume of a quantity of water is 0.42 cm^3. Find its mass in g.

1–7. Find density, dry density, void ratio, water content, porosity and degree of saturation:

(a) Total mass = 89.3 g, mass of water = 21.4 g, total volume = 49.3 cm^3, RD = 2.66

(b) Volume = 1.00 cm^3, mass of water = 0.49 g, dry mass = 1.32 g, RD = 2.70

1–8. A soil sample is 5.08 cm in diameter and 47.3 mm high. Find volume in cm^3. Also find density (g/cm^3) if its mass is 137.33 g.

1–9. Mass of soil + container = 241.13 g.
Mass of container = 106.72 g.
Mass of dry soil plus container = 188.73 g.
Find water content of the soil in usual terms.

1–10. A sample of soil has a mass of 183.4 g. After drying, its mass is 126.2 g. The total volume was measured as 112.2 cm^3. The relative density of the the soil solids is 2.66. Find density, dry density, water content, and void ratio.

1–11. A sample of soil has a mass of 1.693 kg. The water content is 84.4%. Find dry mass.

1–12. Density of a soil is 1.96 g/cm^3, dry density 1.68 g/cm^3, RD 2.65. Find water content, void ratio, porosity and degree of saturation.

1–13. Find dry density, void ratio, and porosity:

(a) Total mass is 101.4 g, total volume is 55.1 cm^3, water content is 28.5% and RD is 2.70.

(b) Density is 2.27 g/cm^3, water content is 8.5% and RD is 2.64.

1–14. The relative density of soil solids for a sand is 2.68. Find the dry density of the sand if it is deposited with a void ratio of 0.90.

1–15. A saturated soil sample (S = 100%) contains 33.4 g of soil solids and 41.6 g of water. The total volume of the sample is 54.1 cm^3. Find water content, density, and relative density of the soil solids.

1–16. Following are the results of a sieve analysis on a granular soil:

Sieve	Retained Mass
12.5 mm (1/2-in)	0 g
9.5 mm (3/8-in)	74.5 g
4.75 mm (No. 4)	217.1 g
2.36 mm (No. 8)	192.3 g
1.18 mm (No. 16)	75.8 g
300 μm (No. 50)	116.9 g
75 μm (No. 200)	83.2 g
Pan	47.4 g

Complete the calculations for grain size distribution curve, and plot the results. Calculate the percentages of gravel, sand, and fines. Calculate the effective size and uniformity coefficient.

1–17. Results of a grain size test are:

Pass 19 mm (3/4-in)	100%
9.5 mm (3/8-in)	82%
2.36 mm (No. 8)	61%
0.60 mm (No. 30)	52%
0.15 mm (No. 100)	36%
0.06 mm	22%
0.01 mm	16%
0.002 mm	11%

Find the percentages of gravel, sand, silt, and clay.

1—18. Results of grain size analysis tests are for three soils, X, Y, and Z.

| Soil X | | | Soil Y | | | Soil Z | |
Size	% Passing		Size	% Passing		Size	% Passing
19 mm (3/4 in)	100		9.5 mm (3/8 in)	100		4.75 mm (No. 4)	100
9.5 mm (3/8 in)	97		4.75 mm (No. 4)	97		0.60 mm (No. 30)	99
2.36 mm (No. 8)	68		1.18 mm (No. 16)	72		0.15 mm (No. 100)	95
0.60 mm (No. 30)	35		0.60 mm (No. 30)	52		0.075 mm (No. 200)	91
0.15 mm (No. 100)	13		0.30 mm (No. 50)	44		0.04 mm	87
0.05 mm	6		0.15 mm (No. 100)	39		0.02 mm	79
0.01 mm	3		0.075 mm (No. 200)	31		0.01 mm	64
			0.05 mm	22		0.005 mm	57
			0.02 mm	6		0.002 mm	46

Plot grain size distribution curves.
Find—for soils X, Y (a) % gravel, sand, fines
$$(b) \ D_{10}, C_u, C_z$$
—for soil Z   (2) % gravel, sand, silt, clay.

1—19. Results of a sieve analysis are:

Sieve	Mass Retained	Find:
9.5 mm (3/8 in.)	0	(a) % gravel
4.75 mm (No. 4)	173 g	(b) % fines (silt and clay)
1.18 mm (No. 16)	104 g	(c) % coarser than 0.30 mm
0.30 mm (No. 50)	216 g	(No. 50).
0.075 mm (No. 200)	97 g	
Pan	32 g	

1—20. A soil has a plastic limit of 27 and a liquid limit of 38. Find the index of plasticity for this soil. Describe the soil. If its natural moisture content is 40%, how would you describe its consistency in the field?

1—21. A soil has a plasticity index of 14.7. It can be rolled to a thread at a water contact of 33.1%. The water content in the field is 46%. Find the Atterberg Limits. Describe condition of the soil in the field.

1—22. Classify soils A to R in figure 1—31
(a) according to Unified System.
(b) according to AASHTO system.

1—23. What are the two main properties measured in laboratory tests to identify and classify soils?

1—24. What tests are used in determining grain size distribution in soils? When is each test used?

GRAIN SIZE DISTRIBUTION (% PASSING EACH SIEVE) AND ATTERBERG LIMITS

Soil / Size	A	B	C	D	E	F	G	H	J	K	L	M	N	P	Q	R
38 mm (1½-in.)	100												100			
19 mm (3/4-in.)	95		100				100			100	100					
9.5 mm (3/8-in.)			85					100		100	100		80			
4.75 mm (No. 4)	86	100	72	100	100	100	87	94		71	51		51			
2.00 mm (No. 10)	71	91	51				72						40			
850 μm (No. 20)	57	70	44	82			55	66	100				29		100	
425 μm (No. 40)	46	20	39	77		73		41		42	29	100	22			
150 μm (No. 100)	33			57			36	18					13	100	82	
75 μm (No. 200)	26	8	32	40	81	56	33	10	89	27	3	66	9	93	70	100
Limits																
w_L	30	N.P.	42	41	61	48	35	N.P.	39	32	N.P.	33	13	N.P.	41	82
I_P	18	N.P.	11	10	36	8	8	N.P.	22	19	N.P.	5	3	N.P.	8	50

FIG. 1–31.

1–25. For Soil A, $w_L = 35$, $w_P = 26$. For Soil B, $w_L = 63$, $w_P = 36$. Samples of the two soils are taken in the field. Their natural moisture contents are 41% for Soil A and 58% for Soil B. Which soil is softer? Why? Describe the consistency of each.

1–26. Find the coefficient of permeability for the soil being tested in the downward-flowing permeameter shown, given that the diameter of the permeameter is 12.5 cm and that 71.3 grams of water flowed through the soil in three minutes.

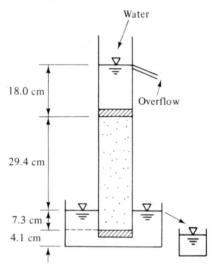

1–27. Water flows through the upward-flowing permeameter shown at the rate of 78.1 g/min. The area of the tube is 20.2 cm². Find the coefficient of permeability. This is a constant head permeameter. (Note that head loss can be measured only between the top of sample and the base of the piezometer tube.)

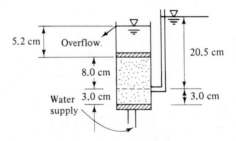

1–28. A clean uniform sand has 10% passing 300 μm (No. 50) sieve. Estimate coefficient of permeability.

1–29. 11.5 g of water flows through this permeameter in 1½ minutes. Tube is

2.75 in. in diameter. Head loss is measured by piezometer tubes in the permeameter as shown. Find the coefficient of permeability of the soil.

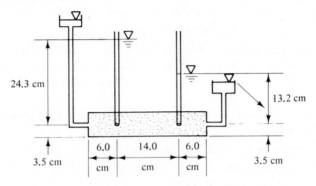

1–30. A uniform soil has 10% passing 425 μm (No. 40) sieve.
   (a) What type of soil is it?
   (b) Estimate coefficient of permeability.
   (c) If the average pore diameter is 1/5 of the effective size, find height of capillary rise.
   (d) Would this soil cause serious frost heaving? Why?

1–31. Following are the results of a sieve analysis:

Sieve	Mass Retained
4.75 mm (No. 4)	0
2.00 mm (No. 10)	41.3 g
425 μm (No. 40)	105.4 g
150 μm (No. 100)	57.1 g
75 μm (No. 200)	10.0 g
Pan	7.3 g

   (a) Calculate and plot the grain size distribution curve.
   (b) Find the percent gravel, sand and fines in the sample.
   (c) Find the effective size and the uniformity coefficient.
   (d) If the average pore size is 1/5 of the effective size, estimate the height of capillary rise in this soil.
   (e) Find the percentage coarser than the 2.00 mm (No. 10) sieve.
   (f) Estimate the coefficient of permeability for this soil using Hazen's formula.

1–32. A falling head permeability test is conducted on a silty soil. Sample diameter is 10.0 cm, and its length or thickness, 7.8 cm. Diameter of the stand pipe is 6.0 mm. The head at the start of the test (see fig. 1–19) is 47.8 cm. The test lasts 8.5 min. and the head falls to 22.6 cm. Find the coefficient of permeability of this soil in cm/s and ft/min.

1–33.   What causes frost heaves? How do ice lenses grow?

1–34.   The average grain size in a sand is 0.2 mm; in a silt, 0.06 mm; and in a clay, 0.004 mm. If the average pore size in each case is one-third of the grain size, estimate the height of capillary rise in each soil in feet and meters.

1–35.   A clean sand has an effective size of 0.13 mm. Estimate the coefficient of permeability.

1–36.   Compare gravels, sands, silts, and clays as to:
   (a)   ease of draining
   (b)   height of capillary rise
   (c)   apparent cohesion

1–37.   What properties contribute to shear strength in (a) clays and (b) sands.

1–38.   List three tests used to measure shear strength of soils in the laboratory.

1–39.   Settlement is usually a more serious problem in clay than in sand. Why?

1–40.   Solve for indicated values:
   (a) $M_W = 315$ kg, $V_W =$ m^3
   (b) $M_D = 4.04$ kg, RD = 2.68, $V_D =$ m^3
   (c) $V_D = 0.613$ m^3, RD = 2.71, $M_D =$ kg
   (d) $V_W = 0.010$ m^3, $M_W =$ kg
   (e) $M_D = 1810$ kg, $V_D = 0.673$ m^3, RD =

1–41.   The volume of a container is 0.050 m^3. It holds 109 kg of dry sand, which has a relative density of 2.65. Find void ratio and density.

1–42.   The density of a soil is 1960 kg/m^3 at a water content of 12.6%. The relative density of the soil is 2.72. Find dry density, porosity, and degree of saturation.

1–43.   The density of a fully saturated organic soil is 1270 kg/m^3. A sample of it with a mass of 161.2 g is dried, and the dry mass is 52.3 g. Find the water content and dry density of the soil, and the relative density of the soil particles.

1–44.   The density of a soil being used in a highway fill is 2020 kg/m^3, and the dry density is 1790 kg/m^3. If the relative density of soil solids is 2.65, find water content, void ratio, and degree of saturation.

1–45.   Soil in a highway is compacted to density of 2130 kg/m^3. A sample of it is tested for water content. Original mass was 163.5 g and dried mass was 140.4 g. Find dry density of highway soil.

1–46. Density of a soil is 1530 kg/m^3. A sample is tested for water content as follows:
- mass soil + beaker 268.41 g
- mass dry soil + beaker 197.94 g
- mass beaker 104.17 g

Find water content and dry density.

1–47. Dry density of a soil in the field is 1570 kg/m^3. Lab tests indicate that its maximum dry density is 1810 kg/m^3, and minimum dry density is 1430 kg/m^3. Find the density index.

1–48. If the relative density of the soil particles in above problem is 2.65, find e, e_{max}, and e_{min}, and check density index calculation.

1–49. Soil in problem 1–47 is densified by a vibrating probe in the field to a density index of 80%. Find the dry density of the soil after compaction. Also calculate its total density if the water content is 11%.

1–50. A cofferdam is built as shown to allow construction along the shore. The dam is 8.5 m wide at the bottom and underlaid by 1.5 m of sand with k = 0.086 cm/s. The head difference is 1.1 m. Find the seepage in (a) m^3/hr per linear meter and (b) m^3/hr for the whole dam. Assume that the length of flow is 8.5 m.

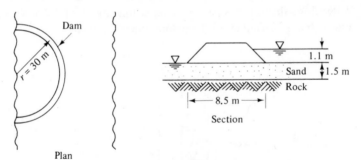

1–51. A 15-cm pipe 12 m in length is clogged with sand with a k of 0.45 cm/s. The head difference is 0.75 m. Find the seepage in m^3/min.

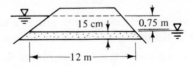

1–52. The dam shown is underlain by a seam of sand 1.1 m thick. The k for sand is .015 cm/s. Find seepage under dam in m^3/s. If dam is 180 m long, find total seepage in m^3/s and m^3/day.

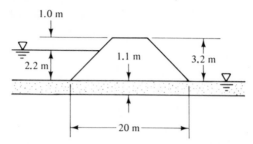

1–53. The canal and river shown are connected by a seam of sand 2.2 m thick ($k = 0.25$ cm/s). Find seepage in m^3/hr per meter of canal and total seepage in m^3/day if the canal is 1.72 km in length.

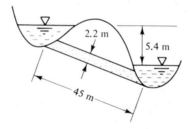

1–54. Water flows through this permeameter at the rate of 31.3 g in 4 minutes. Find k (cm/s). Diameter of the permeameter is 5.0 cm.

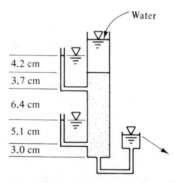

1–55. A soil fails in an unconfined compression test at a load of 259 N. The area of the sample at failure is 10.3 cm^2. Find the shear strength.

1–56. In an unconfined compression test, the sample reached a maximum load of 227 N before failure. The change in length at failure was 0.66 cm. Original diameter and length of sample were 4.50 and 8.81 cm, respectively. Find shear strength.

1–57. An unconfined compression test on a clay sample, 3.80 cm in diameter and 7.25 cm in length, yielded the following readings. The test was terminated at a strain of 1.79 cm.

Strain (cm)	Load (N)
0	0
.15	12.1
.31	21.6
.47	29.8
.63	33.4
.80	41.7
.99	47.2
1.18	53.1
1.33	57.3
1.52	61.7
1.70	65.0
1.79	67.8

Find shear strength (plot load vs strain [%] to obtain values for calculation).

1—58. Strain and load readings taken on a sample in an unconfined compression test are as follows:

Strain (cm)	Load (N)
0	0
0.09	46
0.19	83
0.28	111
0.38	112
0.46	96

Sample was 4.00 cm in diameter and 8.12 cm in length. Plot test results, load vs % strain, find failure load and strain and shear strength.

1—59. In a direct shear test, the total normal force on the sample is 165 N. The shear force required for failure is 104 N. The sample is 10 cm in diameter. Find the normal and shear stresses in kPa.

1—60. A cohesionless sand is tested in a direct shear apparatus with a square shear box 6.00 cm in size. Normal load at failure was 93.8 N with a maximum shear load of 38.7 N. Find stresses on failure plane, sketch a graph showing these on σ, τ axes, and find friction angle and shear strength.

1—61. Three direct shear tests are conducted on a silty clay, with the following results:

Test	Normal Stress (kPa)	Shear Stress (kPa)
1	28	32.4
2	56	36.9
3	84	41.6

Plot the test results: obtain values of c and ϕ for this soil.

1–62. In an unconfined compression test, the following results are recorded:

Initial diameter of sample	35.6 mm
Initial length	68.4 mm
Strain at maximum load	4.9 mm
Maximum load	93.7 N

Find the unconfined compressive strength and the shear strength.

1–63. Solve for indicated values:

(a) $W_W = 31.5$ lb., $V_W =$ ft^3

(b) $W_D = 4.04$ lb., $G_s = 2.68$, $V_D =$ ft^3

(c) $V_D = 0.613$ ft^3, $G_s = 2.71$, $W_D =$ lb

(d) $V_W = 0.010$ ft^3, $W_W =$ lb

(e) $W_D = 118$ lb., $V_D = 0.703$ ft^3, $G_s =$

1–64. The volume of a container is 0.40 ft^3. It holds 43 lb of dry sand, which has a specific gravity of 2.65. Find void ratio and unit weight.

1–65. The unit weight of a soil is 122 lb/ft^3 at a water content of 12.6%. The specific gravity of the soil is 2.72. Find dry unit weight, porosity, and degree of saturation.

1–66. The unit weight of a fully saturated organic soil is 79 lb/ft^3. A sample of it with a mass of 161.2 g is dried, and the dry mass is 52.3 g. Find the water content and dry unit weight of the soil, and the specific gravity of the soil particles.

1–67. The unit weight of a soil being used in a highway fill is 126 lb/ft^3, and the dry unit weight is 112 lb/ft^3. If the specific gravity of soil solids is 2.65, find water content, void ratio, and degree of saturation.

1–68. Soil in a highway is compacted to unit weight of 133 lb/ft^3. A sample of it is tested for water content. Original mass was 163.5 g and dried mass was 140.4 g. Find dry density of highway soil.

1–69. Unit weight of a soil is 114 lb/ft^3. A sample is tested for water content as follows:

weight of soil + beaker 228.41 g
weight of dry soil + beaker 197.94 g
weight of beaker 104.17 g

Find water content and dry unit weight.

1–70. Dry unit weight of a soil in the field is 98 lb/ft^3. Lab tests indicate that its maximum dry unit weight is 113 lb/ft^3 and minimum dry unit weight is 89 lb/ft^3. Find the relative density.

1–71. If the specific gravity of the soil particles in above problem is 2.65, find e, e_{max}, and e_{min}, and check relative density calculation.

1–72. Soil in problem 1–70 is densified by a vibrating probe in the field to a relative density of 80%. Find the dry unit weight of the soil after compaction. Also calculate its total unit weight if the water content is 11%.

1–73. A cofferdam is built as shown to allow construction along the shore. The dam is 28 ft wide at the bottom and underlaid by 3.0 ft of sand with $k = 0.0029$ ft/s. The head difference is 3.5 ft. Find the seepage in (a) ft^3/hr per linear foot and (b) gallons/hr. for the whole dam. Assume that the length of flow path is 28 ft.

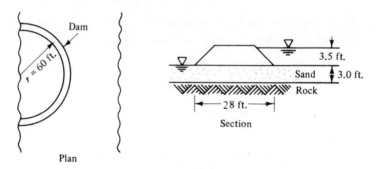

1–74. A 6 in pipe 30 ft in length is clogged with sand with a k of 0.020 ft/s. The head difference is 2.2 ft. Find the seepage in ft^3/min.

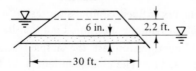

1–75. The dam shown is underlain by a seam of sand 1.3 ft thick. The k for sand is 5.0×10^{-4} ft/s. Find seepage under dam in ft^3/s. If dam is 520 ft long, find total seepage in ft^3/s and ft^3/day.

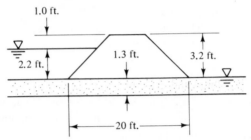

1–76. The canal and river shown are connected by a seam of sand 6.5 ft thick (k = 0.50 ft/min.). Find seepage in ft^3/h per foot of canal and total seepage in ft^3/day if the canal is 1.72 miles in length.

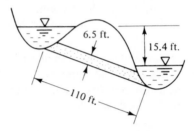

1–77. Water flows through this permeameter at the rate of 31.3 g in 4 minutes. Find k (cm/s). Diameter of the permeameter is 2.0 in.

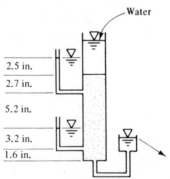

1–78. A soil fails in an unconfined compression test at a load of 70 lb. The area of the sample at failure is 1.60 in^2. Find the shear strength.

1–79. In an unconfined compression test, the sample reached a maximum load of 53.5 lb before failure. The change in length, at failure, was 0.26 in. Original diameter and length of sample were 1.75 and 3.62 in, respectively. Find shear strength.

1–80. An unconfined compression test on a clay sample, 1.85 inches in diameter and 3.88 inches in length, yielded the following readings. The test was terminated at a strain of 0.90 in.

Strain (in)	Load (lb)
0	0
.08	12.1
.15	21.6
.24	29.8
.31	33.4
.40	41.7
.49	47.2
.59	53.1
.67	57.3
.76	61.7
.85	65.0
.90	67.8

Find shear strength (plot load vs strain [%] to obtain values for calculation).

1−81. Strain and load readings taken on a sample in an unconfined compression test are as follows:

Strain (in)	Load (lb)
0	0
0.09	46
0.19	83
0.28	111
0.38	112
0.46	96

Sample was 1.90 inches in diameter and 4.02 inches in length. Plot test results, load vs % strain, find failure load and strain and find shear strength.

1−82. In a direct shear test, the total normal force on the sample is 41.3 lb. The shear force required for failure is 26.0 lb. The sample is 2.50 inches in diameter. Find the normal and shear stresses in lb./ft.^{2} kPa.

1−83. A cohesionless sand is tested in a direct shear apparatus with a square shear box 2.0 inches in size. Normal load at failure was 19.7 lb with a maximum shear load of 10.3 lb. Find stresses on failure plane, sketch a graph showing these on σ, τ axes, and find friction angle and shear strength.

1−84. Three direct shear tests are conducted on a silty clay, with the following results:

Test	Normal Stress (lb/in²)	Shear Stress (lb/in²)
1	28.0	32.4
2	56.0	36.9
3	84.0	41.6

Plot the test results: obtain values of c and ϕ for this soil.

1–85. In an unconfined compression test, the following results are recorded:

Initial diameter of sample	1.40 in
Initial length	2.86 in
Strain at maximum load	0.102 in
Maximum load	26.5 lb

Find the unconfined compressive strength and the shear strength.

2

Soils Investigation

In the evaluation of an area for construction of buildings or other structures, or as a source of construction material, the soil conditions must be investigated before any detailed designs are made.

A soils investigation involves field sampling and testing, laboratory analysis, and preparation of a report. The planning and evaluation of the field work are aided by a knowledge of the mechanics of soil deposits' formation.

2–1 SOIL DEPOSITS

As discussed in chapter 1, soils result from weathering of the parent material—rock—that forms the earth's crust. These soils are then often transported, sorted, and deposited in various formations.

2–1.1 Soil grains result from weathering of bedrock. As noted in chapter 1, physical weathering produces granular soils, gravel, sand, and silt (depending on the size of the particles); chemical weathering creates the very small clay grains. Grains are often transported from their original location by water. Since the size particle that can be carried by running water depends on the velocity of the water, soil grains are sorted and deposited in layers that correspond with their size. As a stream slows down, gravels are deposited first, followed by sands. Silts are left at the mouths of streams. The still waters of lakes and seas allow clay grains to settle out, thus forming clay plains.

2—1.2 Soil deposits are either *residual* or *transported.*

Residual soils are the product of weathering of the existing bedrock. The depth of weathering is greatest in hot, humid climates; it may be several hundred feet in the tropics.

Transported soils, as the name suggests, have been moved from their place of origin. The main transportation agents are:

1. *Rivers and streams.* Material carried by the water is sorted. Gravels, sands, and silts are deposited in that order as the speed of the river or stream slows.
2. *Lakes.* Clays and silts settle out in calm lake water.
3. *Wind.* Sand dunes are formed from wind-carried fine sands, and loess deposits are silts.
4. *Glaciers.* As glaciers moved across Canada and the northern part of the United States, they eroded, transported, and deposited soils in many types of formations. (See fig. 2—1.)

Glaciers advanced and retreated many times during the last Ice Age. As they advanced, they carried or shoved a vast amount of soil. When they stopped or melted, large mounds or plains of mixed soil, gravel, sand, silt, and clay were left. This soil is called *till.* Other deposits were carried by the vast quantities of melt water associated with the glaciers; these deposits were sorted and deposited as sands, silts or clays in glacially formed streams or lakes. Many glacial lakes deposited alternate layers of fine silt and clays, with silts settling in the summer and clays in the winter when the lake was calm or frozen over. These deposits are called *varved clays,* and are composed of alternate layers 5 mm to 10 mm (1/4 in to 1/2 in) thick. Part of the land area of North America was depressed by the glaciers and flooded with salt water from the ocean. Clay deposited in this salt water formed a very soft, low-density deposit called *marine clay.*

2—1.3 Types and characteristics of transported soil deposits are summarized in table 2—1.

2—2   FIELD INVESTIGATION TECHNIQUES

Soils investigations are conducted for most medium- to large-sized buildings, highways, bridges, dams, water control facilities, harbors, and other structures. The purpose is to find the allowable bearing capacity for foundations. Investigations are also conducted to determine water resources, to find aggregates, to estimate infiltration and seepage rates, and to help assess land use capabilities.

FIG. 2–1. Glaciation of North America (David F. McCarthy, *Essentials of Soil Mechanics and Foundations*, 1977. Reprinted with permission of Reston Publishing Company, Inc., A Prentice-Hall Company, 11480 Sunset Hills Rd., Reston, VA 22090).

2–2.1 Information usually required in soils investigations includes:

— depth, thickness, and properties of each soil layer
— location of groundwater table
— depth to bedrock

2–2.2 Before a field investigation is carried out at the site, preliminary information regarding soil condition can often be obtained from the following sources:

Table 2–1

MAIN TYPES OF SOIL DEPOSITS

Type	Formation	Characteristic	Material	Significance
Glacial				
End or lateral moraine	At front or sides of glacier	Rough, rolling ground	Till-mixed	Hard, fairly impervious
Ground moraine	Under glacier	Rolling	Till-mixed	Hard, fairly impervious
Drumlin	Under glacier	Mounds about 1 km × 1/2 km (1/2 mi × 1/4 mi)	Till-mixed	Hard, fairly impervious
Kame moraine	Water flowing over or out of glacier	Rough, rolling ground	Sand, gravel, silt	Aggregates source
Esker	Streams flowing under the glacier	Long ridges	Sand, gravel	Aggregates source
Outwash or spillway	Streams flowing away from glaciers	Fan-shaped delta	Sand, gravel, silt	Aggregates source
Lake bottom	Soil grains settling out	Flat	Clays, varved clays	Poor foundation soil, soft and compressible
Beach	Shores, bars, etc., of lakes	Beach-type, sorted deposits	Sand, gravel	Aggregates source
Marine	Soil grains settling out in salt water	Flat	Marine clays	Very soft, very sensitive and compressible
Post-glacial				
Alluvial	Stream	Existing stream valley	Sand, gravel, silt	May be aggregates or pockets of soft, variable material
Dune, loess	Wind	Small hills	Sand, silt	Uniform particles
Beach	Shores of lakes, oceans	Beach, well-sorted	Sand, gravel, uniform	Aggregates source
Organic	Low, poorly drained areas	Marshes, muskegs	Peat, muskeg, muck	Very weak, extremely high compressibility

1. *Geological and agricultural soils maps.* These often indicate the types of soil or geological formation that cover the area being investigated.
2. *Aerial photographs.* Drainage patterns can be identified, and color and tone of photos give a good indication of the type of soil that might be encountered.
3. *Area reconnaissance.* The condition of other buildings in the area can give some clue as to potential foundation problems. The depth to water level in adjacent wells may indicate the elevation of the groundwater table.

2–2.3 Subsurface investigation of soils deposits can be carried out by five main methods.

1. *Geophysical methods (seismic or electrical).* Variations in the speed of sound waves or in the electrical resistivity of various soils are especially useful indicators of the depth to the water table and to bedrock. Some typical seismic wave velocities are shown in table 2–2.
2. *Probing or jetting with a stream of water.* In this method the material washed up and left at the surface does not represent the soil found since the fines are washed away. Also, it is difficult to establish the depths at which various layers are encountered.
3. *Test pits cr trenches.* This method is suitable for shallow depths only.
4. *Hand augers.* Again, this method is suitable for shallow depths only. Only disturbed or mixed samples of soil can be obtained.
5. *Boring test holes and sampling with drill rigs.* This is the principal method for detailed soils investigations, and is described in detail in the following sections.

Table 2–2

SEISMIC WAVE VELOCITIES

Material	Velocity	
	(m/s)	*(ft/s)*
Loose, dry sand	150–450	500–1500
Hard clay, partially saturated	600–1200	2000–4000
Loose saturated soil	1400–1800	5000–6000
Saturated soil	1200–3000	4000–10000
Weathered rock	1200–3000	4000–10000
Sound rock	2000–6000	7000–20000

2–2.4 The number and depth of test holes depend on the structure to be built, the type of soil, and the variation in the soil profile found. Following are typical requirements.

1. *One-story buildings.* Test holes are drilled 30-60 m (100-200 ft) apart to a depth of 6-10 m (20-30 ft), with one deeper hole to check for weaker soils.
2. *Four-story buildings.* Test holes are drilled 15-30 m (50-100 ft) apart to a depth of 10-15 m (30-50 ft), with some holes to a depth one and one-half times the width of the building, and at least one deep hole to bedrock.
3. *Highways.* Test holes are drilled 300 m (1000 ft) apart to a depth of 3 m (10 ft) below subgrade level.

Samples and field tests (especially the first few holes) are taken in every soil layer or every 1.5 m (5 ft), whichever is less.

2–2.5 Test holes are opened with a continuous flight auger which brings the soil up to the surface. The most common size of auger is 10 cm (4 in). The auger is rotated by a drilling machine mounted on a truck or tracked vehicle. The auger is removed to insert sampling tools into the test hole. In some cases, especially where there are granular soils below the water table, the hole does not remain open when the auger is removed. Under such conditions, either (1) the hole is cased, that is, a pipe is driven in and the soil is augered out inside the pipe to the bottom, or (2) a hollow-stem auger is used, which allows sampling tools to be put down the hollow stem of the auger to the bottom for sampling. A drill rig is shown in fig. 2–2; some drilling equipment, in fig. 2–3.

2–2.6 Samples taken during the soils investigation may be *undisturbed* or *disturbed.* In undisturbed samples the structure of the soil in the sample is as close as possible to the structure of the soil in the field.
    The main types of samples taken are:

1. *Auger samples.* The sample taken from the soil is brought up by the auger (depth is not certain); the sample is disturbed.
2. *Split spoon.* The spoon is lowered to the bottom of the hole, attached to the drill rods, and driven into the soil; the sample is disturbed; this technique is used in all soils.
3. *Thin-wall sampler (Shelby tube).* The sampler is attached to drill rods, lowered to the bottom of the hole, and pushed into the soil in one smooth motion; the sample is waxed when removed; the sample is undisturbed. This technique is used only in clays and silts.
4. *Rock cores.* Samples are taken with diamond drill bits.

FIG. 2–2. Soils investigation drill rig: (a) drill rig with the standard
penetration hammer attached to the cable; (b) augering to open a
test hole; (c) split spoon sampler attached to drill rods has been
lowered to the bottom of the test hole through the hollow stem of
this auger and chalk marks have been placed on the rod at 15 cm. (6
in.) intervals; (d) drill rod and spoon being driven into soil at the
bottom of the test hole.

(a)

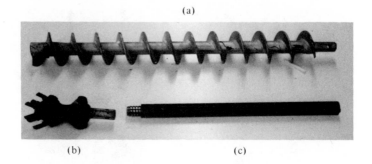

(b)   (c)

FIG. 2–3. Drilling equipment: (a) a section of continuous flight auger; (b) auger head used to loosen the soil; (c) drill rod used to lower sampling tools into the test hole.

Soil samples are put into containers, closed to prevent evaporation, and labelled.

2–2.7   The methods used in field or in-place testing are:

1. *Standard penetration test.* The sampler, 60 cm (2 ft) long with a 50-mm (2-in) outside diameter, is driven by a hammer with a mass of 63.5 kg (140 lb) and falling 75 cm (30 in). The sampler is first driven 15 cm (6 in) to be sure that it is below the bottom of the test hole, and then the number of blows required to drive it another 30 cm (12 in) is recorded as the N value. This test is the most common strength test conducted in the field. It is used with all soils except gravels, and is often used directly for the design of foundations on granular soils. Descriptive terms for soil conditions measured by this test are listed in table 2–3. A soil sample (disturbed) is also obtained in the spoon.
2. *Vane*—the vane is shoved into soil and torque applied until it twists. This gives the shear strength, or cohesion of cohesive soils.
3. *Cone*—the cone is driven through soils, with the number of blows required for each foot being recorded. This indicates the depth of fill or the depth to layer changes.

The equipment used in these tests is illustrated in fig. 2–4.

2–2.8   The *pressuremeter test,* widely used in Europe, is a much more accurate and scientific field strength test than the standard penetration test. The test is now being introduced in many soils investigations in North America.

  It consists basically of a probe that is lowered into a test hole to the desired depth, a water-filled volumeter, and a pressure source, usually com-

Table 2–3

FIELD TERMS TO DESCRIBE SOIL CONDITIONS, BASED ON
THE STANDARD PENETRATION TEST

N = Blows/30 cm (1 ft)	*Relative Condition of Sand and Silt Soils*
0–4	Very loose
5–10	Loose
11–30	Medium dense
31–50	Dense
more than 50	Very dense
	Consistency of Clays
0–1	Very soft
2–4	Soft
5–8	Firm
9–15	Stiff
16–30	Very stiff
more than 30	Hard

(3)

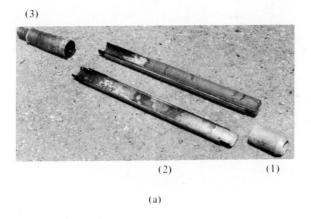

(2)   (1)

(a)

(b)   (c)

FIG. 2–4. Equipment used in soil testing in the field: (a) split spoon showing (1) cutting shoe, (2) sample in the opened split spoon, and (3) head to attach sampler to the drill rod; (b) cone; and (c) vane, 1½ in. in diameter and 3 in. long.

pressed gas (see fig. 2–5). The probe consists of three cells that are covered with a rubber membrane, allowing them to expand under pressure. The outer cells, or guard cells, ensure that the measuring cell cannot expand in the vertical direction as pressure is applied.

After installation of the probe in the test hole, pressure is applied to the guard cells and to the water in the volumeter in about ten increments. Readings at each pressure increment are taken of the volume of water in the volumeter, which indicates the increase in volume of the measuring cell as pressure increases.

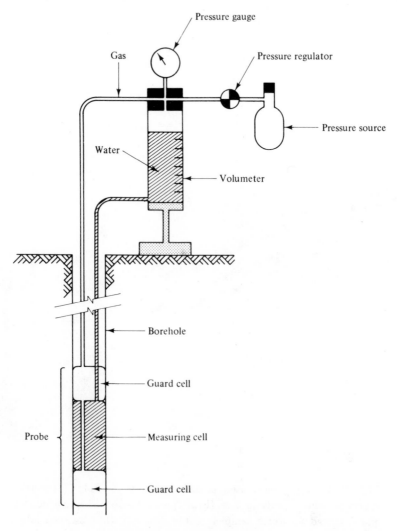

FIG. 2–5. The pressuremeter.

Values of pressure-vs-volume change can be plotted to give reliable measurements in predicting settlement and shear strength properties of the soil.

2–2.9 A very important part of the soils investigation is to establish the water table elevation. This is done by measuring down the hole to the final water surface as water fills the hole.

In granular soils, water table elevation is easy to determine since water flows in quickly and fills the hole. Usually the walls of the hole cave in up to the water table level. In clay soils or soils mixed with clay, a long time may elapse before a sufficient quantity of water seeps out of the soil to fill the hole. Judgement is involved in assessing the significance of water level measurements in these soils.

In many cases where the water table cannot initially be determined or when the change in water table elevation during the year or during construction is required, a piezometer is installed. This usually is a porous cylinder filled with coarse sand. A tube leads to the surface. It is installed in the hole, and partially backfilled with sand. The hole is then sealed with bentonite, which swells when water reaches it, thus preventing surface water from affecting the piezometer. The rest of the hole is then backfilled with soil. Water will fill the tube to the level of the groundwater table, and this level rises or falls with the water table in the surrounding soil. A two-conductor wire, connected to a battery and a voltmeter, is inserted in the tube. When the wire touches the water surface, a circuit is formed and indicated on the meter. Therefore, the depth to the water surface and the water table can be established at any time. A piezometer installation is shown in fig. 2–6.

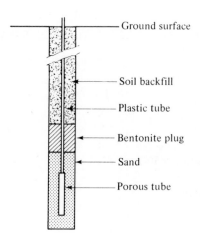

FIG. 2–6.   Piezometer installation.

2–2.10   As the soils investigation is conducted, a field log of the test hole must be kept. This log should include:

1. Sample number, depth, and type.
2. Field tests, depth, and results.
3. Depth to layer changes.
4. Field soil description.
    a. Type of soil grains.
    b. Moisture conditions.
    c. Consistency or density.
    d. Seams and stratification.
    e. Other distinguishing features.

Typical test hole log notes are shown in fig. 2–7.

Identification criteria to help describe soils according to the Unified System are given in fig. 2–8.

2–3 LABORATORY TESTING

The next step in a soils investigation is the testing of the samples.

2–3.1   Representative samples of each soil type found at the site are selected for initial testing. Further tests on other samples may also be required. The most common types of tests are listed in table 2–4.

Table 2–4

LABORATORY TESTS RELATED TO A SOILS INVESTIGATION

Test	Sample Required		Soils	
	Disturbed or undisturbed	Undisturbed	Cohesive	Granular
Moisture content	X		X	X
Grain size	X		X	X
Atterberg Limits	X		X	
Relative density (specific gravity)	X		X	X
Density (unit weight)		X	X	X
Unconfined compression		X	X	
Triaxial compression		X	X	X
Direct shear		X	X	X
Consolidation		X	X	
Vane shear		X	X	
Permeability		X		X

Test Hole No.	1

Date	82/6/1

Drilling Method	Auger

Depth to water:

immediate	1.2 m

24 hours	1.0 m

Depth (m)	Soil Description	Sample			SPT Value "N"
		No.	Type	Depth (m)	
0 —					
	Topsoil, sandy (0-0.3 m)				
0.3 —					
0.5 —					
	Loose brown moist fine sand (0.3-1.2 m)	1	Auger	0.8	
1.0 —					
1.2 —					
1.5 —	Dense silty sand 1.2-1.8m	2	Split Spoon	1.2-1.6	8/12 - 20
1.8 —					
2.0 —					
	Till, clayey, with some silt, moist, hard (1.8-2.8m)	3	Split Spoon	2.1 -2.5	15/18 = 33
2.5 —					
2.8 —					
3.0 —					
	Clay, soft wet (2.8-3.6 m)	4	Shelby	3.0-3.6	
3.5 —					
3.6 —					
	End of test hole				

FIG. 2–7.   Typical field notes.

2–3.2   Approximate values for soil strength may be obtained from simple field tests, as indicated in table 2–5.

2–4   SOIL REPORTS

The final step in a soils investigation is the preparation of a soils report. This report includes a summary of the test program, a general description of the soil conditions, a detailed analysis of each type of soil found, and recommendations for design (as required). A copy of the test-hole logs and the soil profile is also included, and these are the only parts of the report discussed here.

2–4.1   The test-hole logs summarize the field and laboratory information gained about each test hole. Figure 2–9 contains typical symbols used to draw a test-hole log. Their use is demonstrated in example 2–1.

Coarse grained soil

Gravel – more gravel than sand		Sand – more sand than gravel	
Clean less than 5% pass the No. 200 GW or GP	Dirty over 12% pass the No. 200 GC or GM	Clean less than 5% pass the No. 200 SW or SP	Dirty over 12% pass the No. 200 SC or SM

Note – well graded – (W) wide range in grain sizes.

– poorly graded (P) – one size range.

– clayey (C) (see opposite to tell difference between clayey and

– silty (M) silty fines.)

Fine grained soil

Identification (see also Table 1.2)

	Low compressibility CL or ML		High compressibility CH or MH	
	Dry strength (crushing characteristics)	Dilatancy (reaction to shaking)	Toughness (consistency near plastic limit)	
	None to slight	Quick to slow	None	ML
	Medium to high	None to very slow	Medium	CL
	Slight to medium	Slow to none	Slight to medium	MH
	High to very high	None	High	CH

FIG. 2–8. Identification criteria for describing soils according to the Unified System.

Table 2−5

APPROXIMATE VALUES FOR SOIL STRENGTH

Densities of Granular Soils

Description	Density Index	Approx. ϕ	Field Identification
Very loose	0%–15%	$<28°$	Easily penetrated by a wooden survey stake
Loose	15%–35%	$28°$-$30°$	Easily penetrated by a reinforcing rod pushed by hand
Medium dense	35%–65%	$30°$-$36°$	Easily penetrated by a reinforcing rod driven with a hammer
Dense	65%–85%	$36°$-$40°$	Penetrated 25–50 cm by a reinforcing rod driven with a hammer
Very dense	85%–100%	$>40°$	Penetrated only a few centimeters by a reinforcing rod driven with a hammer

Consistencies of Cohesive Soils

Consistency	Field Identification	Approximate Shear Strength kPa	lb/ft^2
Very soft	Easily penetrated several centimeters by the fist	<12	<250
Soft	Easily penetrated several centimeters by the thumb	12–25	250–500
Firm	Can be penetrated several centimeters by the thumb with moderate effort	25–50	500–1000
Stiff	Readily indented by the thumb but penetrated only with great effort	50–100	1000–2000
Very stiff	Readily indented by the thumbnail	100–200	2000–4000
Hard	Indented with difficulty by the thumbnail	>200	>4000

2−4.2 A soils profile is a drawing that shows two or more test holes in elevation, and indicates where each soil type was found. A typical profile is also included in example 2−1.

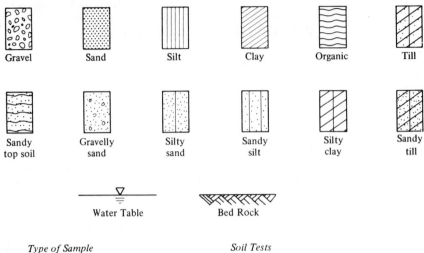

Gravel	Sand	Silt
Clay	Organic	Till
Sandy top soil	Gravelly sand	Silty sand
Sandy silt	Silty clay	Sandy till

Water Table   Bed Rock

Type of Sample

S.S. – Split Spoon
S.T. – Shelby Tube
A.S. – Auger Sample
W.S. – Washed Sample
R.C. – Rock Core
V.S. – Vane Shear Sample

Soil Tests

F.V. – Field Vane
L.V. – Lab Vane
Qu – Unconfined Compression
Qq – Undrained (quick) Triaxial
C – Dynamic Cone (Blows per foot)

FIG. 2–9.   Typical test hole symbols and abbreviations.

Example 2–1:   Following are results of a soils investigation:

1. TEST HOLE LOCATIONS

Plan

Elevations

Hole No. 1   575.5 m
2   574.7 m
3   576.2 m

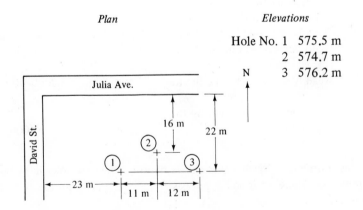

2. FIELD NOTES

Test Hole Logs

Hole No.	Depth (m)	Description
1	0-0.4	Topsoil
	0.4-1.4	Gray, silty clay till—moist
	1.4-2.0	Clay, some silt, seams of sand—wet
	2.0-3.8	Silty sand—saturated
2	0-0.3	Topsoil
	0.3-0.8	Gray, silty clay till—moist
	0.8-1.1	Coarse sand, some gravel
	1.1-2.4	Clay, some silt, seams of sand—wet
	2.4-5.2	Silty sand—saturated
	5.2	Rock
3	0-0.4	Topsoil
	0.4-1.2	Gray, silty clay till—moist
	1.2-1.8	Clay, some silt—wet
	1.8-2.4	Brown, silty till, some sand and clay—wet
	2.4-4.4	Silty sand—saturated

Field Samples and Tests

Sample No.	Hole	Depth (m)	Type	N
1	1	0.5-0.9	S.S.	12
2	1	1.6-2.0	S.S.	4
3	1	2.5-2.9	S.S.	21
4	1	3.4-3.8	S.S.	33
5	2	0.5-0.9	S.S.	8
6	2	1.5-1.9	S.S.	3
7	2	2.0-2.3	S.T.	—
8	2	3.0-3.4	S.S.	34
9	2	4.0-4.4	S.S.	60
10	3	0.5-0.9	S.S.	14
11	3	1.2-1.6	S.T.	—
12	3	1.9-2.3	S.S.	25
13	3	3.3-3.7	S.S.	40

Vane shear tests were conducted in holes 1 and 3:

Hole	Depth (m)	Cohesion
1	1.5	30 kPa (600 lb/ft^2)
3	1.7	33 kPa (660 lb/ft^2)

TEST HOLE LOG

Hole No. _____ 1 _____   Site _____ Julia Ave. & David St. _____

Date drilled _____ 82 – 08 – 07 _____   Elevation _____ 575.5 _____

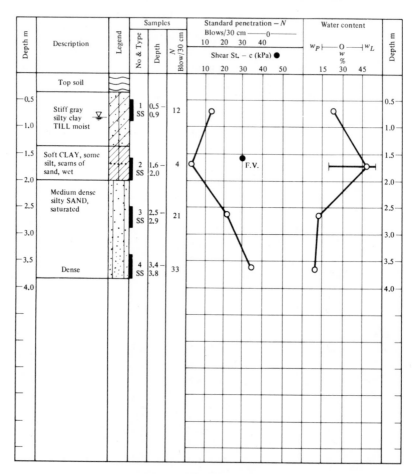

FIG. 2–10. Test hole log.

Water levels one day after the holes are drilled:

Hole No. 1	0.8 m
Hole No. 2	0.3 m
Hole No. 3	1.5 m

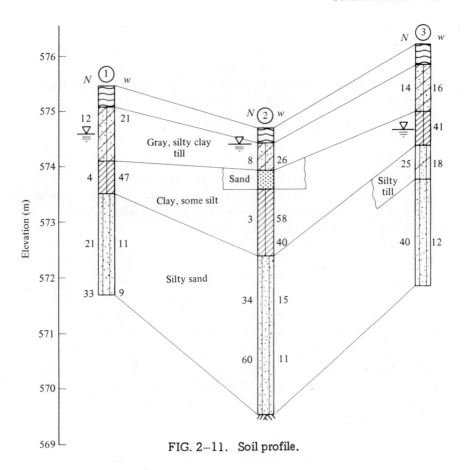

FIG. 2–11. Soil profile.

3. LABORATORY TESTS

Sample No.	w	w_L	w_P	Shear Strength kPa (lb/ft²) (Unconfined Compression Test)
1	21	–	–	–
2	47	53	21	–
3	11	–	–	–
4	9	–	–	–
5	26	–	–	–
6	58	55	20	–
7	40	51	26	42 (850)
8	15	–	–	–
9	11	–	–	–
10	16	–	–	–
11	41	58	29	65 (1300)
12	18	–	–	–
13	12	–	–	–

4. TEST HOLE LOG

5. SOIL PROFILE

2–5 PROBLEMS

2–1. What is a till soil? In which formations is it found?

2–2. List four soil formations that might contain sand and gravel deposits.

2–3. How would you recognize an esker?

2–4. What is varved clay? How is it formed?

2–5. Where are clays usually deposited? Why?

2–6. How does water sort soils?

2–7. What information is required from a soils investigation?

2–8. A building, five stories in height and 25 m by 120 m (80 ft × 400 ft) in size, is being designed. What type of soils investigation (number and depth of holes) would you recommend?

2–9. How are test holes usually opened? In some soils, the hole will probably fill in after the augers are withdrawn. In what type of soil would you expect this problem? What should be done in this case?

2–10. Describe how undisturbed samples are taken in a soils investigation.

2–11. Describe the types of samples taken in a soils investigation.

2–12. List the field tests conducted during a soils investigation, and indicate what soil they are used for.

2–13. What precautions must be taken when locating the groundwater table by measuring the depth to the water level in test holes? Why?

2–14. What is the standard penetration test? How is the N value found?

2–15. What is a drumlin?

2–16. What is a piezometer? For what is it used?

2–17. A drill hole usually indicates that soils have been deposited in layers. Why are they deposited this way?

2–18. Explain the basic principles of the seismic method of soils investigation.

2–19. At what depth intervals should samples be taken in a drill hole?

2–20. The standard penetration value in a sand is 8. The _____ of the sand is _____ .

2–21. The standard penetration value in a clay is 6. The _____ of the clay is _____ .

2–22. List eight tests conducted on soils from a soils investigation. Indicate (a) whether each requires an undisturbed sample, and (b) whether each is used for granular or clay soils.

2–23. Estimate the shear strength (cohesion) of a clay described as soft.

2–24. A reinforcing rod can be driven about 30 cm into a sand soil. Estimate the soil's strength and describe its condition.

2–25. What information is usually shown on a test hole log?

2–26. Draw test hole logs for holes 2 and 3 in example 2–1.

3

Compaction

In using soil as a material in highway embankments, earth dams, and backfill for various types of construction, the quality of the earth construction is controlled mainly by *compaction* requirements. The compaction test is the most common field test for soils during construction.

3-1 MAXIMUM DRY DENSITY

Compaction requirements are measured in terms of the dry density of the soil. The expected value for dry density varies with the type of soil being compacted. For example, a clay soil may be rolled many times and not reach 2000 kg/m^3, whereas a granular soil may have a dry density above this value without *any* compactive effort. Therefore a value for the maximum dry density that can be expected must be established for each soil.

3-1.1 Under any compactive effort, the dry density of a soil varies with its water content. A soil compacted dry will reach a certain dry density. If compacted again with the same compactive effort—but this time with water in the soil—the dry density will be higher, since the water lubricates the grains and allows them to slide into a denser structure. Air is forced out of the soil, leaving more space for the soil solids, as well as the added water. With an even higher water content, a still greater dry density may be reached since more air is expelled. However, when most of the air in the mixture has been removed, adding more water to the mixture before compaction results in a lower dry density—the extra water merely takes the place of some of the soil solids. This principle is illustrated in fig. 3—1.

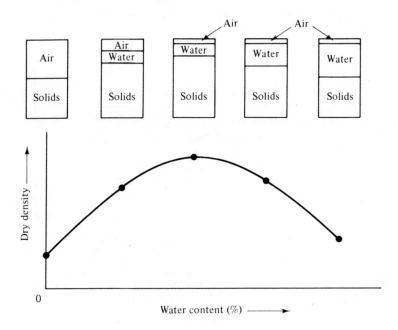

FIG. 3—1.   Variation of dry density with water content.

3—1.2 The first step in compaction control is finding (1) the maximum dry density that can be expected for a soil under a certain compactive effort, and (2) the water content at which this density is reached. These are obtained from a *compaction curve,* as shown in fig. 3—2. The compaction curve is also called a *moisture–density curve* or a *Proctor curve* (named after the originator of the test). The curve is plotted from the results of the compaction test (moisture-density test or Proctor test), which is described in the next section. Dry density is plotted against water content, and a curve is drawn through the test points. The top of the curve represents (1) the maximum dry density for the soil with the test compactive effort, and (2) the corresponding water content which is called the *optimum water content* (w_o).

3—1.3 In the standard compaction test, the soil is compacted in a 10-cm (4-in) diameter mold having a volume of 943.9 cm^3 (1/30 ft^3). The soil is placed in three layers, each compacted with 25 blows of a 2.5-kg (5.5-lb) hammer falling

Compaction test results:

Trial No.	1	2	3	4	5	6
Dry density (kg/m^3)	1862	1894	1927	1929	1883	1836
Dry unit weight (lb/ft^3)	(116.2)	(118.2)	(120.2)	(120.4)	(117.5)	(114.6)
Water content (%)	6.8	9.1	11.0	12.8	15.0	16.6

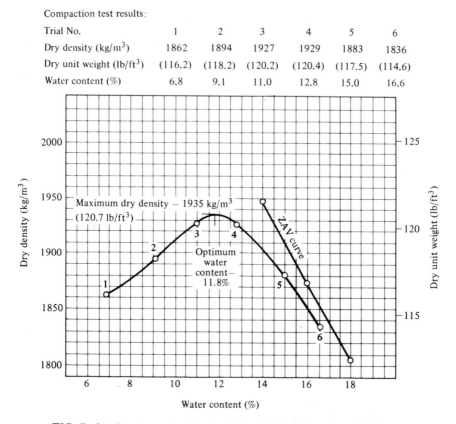

FIG. 3–2. Compaction curve. (See section 3–1.4 for a description of the ZAV curve.)

30 cm (12 in). (See fig. 3–3.) Usually the test is conducted on only that material that passes the 4.75-mm (No. 4) sieve. If coarser grains are to be included in the test, a larger mold—15 cm (6 in) in diameter with a volume of 2124 cm^3 (0.075 ft^3)—is used. With these coarse materials, the number of blows per layer is increased to 56 (from 25) to maintain the same compactive effort. This compactive effort was originally chosen by Proctor to approximate the effort provided by construction rollers at the time (1930s).

In the test, a sample of the soil is mixed with water, and then compacted. The mass of the compacted sample is measured, and part of it is taken to dry for the purpose of determining water content. More water is then added to the soil, and it is compacted again. This procedure is repeated until the density decreases. The calculations involved are illustrated in example 3–1.

Example 3–1: The following results were obtained in trial 1 of a standard compaction test.

Density sample
    Mass of soil plus mold   = 5321 g
    Mass of mold   = 3449 g
Water content sample
    Mass of sample plus container   = 252.60 g
    Mass of dry sample plus container = 238.04 g
    Mass of container   = 104.73 g
Calculations: Water content sample
    Mass of water   = 14.56 g
    Mass of dry soil   = 133.31 g
    Water content (w)   = 10.9%
From density sample
    Mass of soil in mold   = 1872 g (4.127 lb)
    Density (1872 g/943.9 cm^3)* = 1.983 g/cm^3 (123.8 lb/ft^3)
                                    = 1983 kg/m^3
    Dry density (1983 kg/m^3/
        1.109)**   = 1788 kg/m^3 (111.6 lb/ft^3)

  * Volume of standard mold is 943.9 cm^3 (1/30 ft^3)
 ** Dry density = density/(1 + w)

One point of the moisture-density curve is 1788 kg/m^3 at 10.9% water.

3–1.4 As an aid in drawing the moisture-density curve and as an indication of the maximum theoretically possible density, the *zero air voids (ZAV)* curve is often plotted. This curve joins points giving the maximum theoretical density of

FIG. 3–3. Standard compaction test apparatus—the compaction mold, collar, and compaction hammer.

the soil at various moisture contents, that is, with no air left in the soil-water mixture. Points on this curve can be obtained with this equation:

$$\text{ZAV } \rho_D = \frac{\rho_w}{\dfrac{1}{RD} + w} \qquad (3-1)$$

Example 3–2: Given that RD (G_s) = 2.72, find ZAV ρ_D when w = 10%.

$$\text{ZAV } \rho_D = \frac{1000 \text{ kg/m}^3}{\dfrac{1}{2.72} + 0.10} = 2138 \text{ kg/m}^3$$

or

$$\text{ZAV } \gamma_D = \frac{62.4 \text{ lb/ft}^3}{\dfrac{1}{2.72} + 0.10} = 133.4 \text{ lb/ft}^3$$

The ZAV curve for a soil with RD = 2.68 is shown in fig. 3–2.
   The ZAV curve helps to establish the compaction curve for a soil, as follows:

1. No point can be above the ZAV line. Therefore errors are obvious.
2. The slope of the moisture-density curve on the wet side of optimum moisture content is parallel to the ZAV curve. This is especially helpful in sketching the curve where test results are erratic, as is commonly the case with sands.

3–1.5 The moisture-density curve is different for each soil. Granular, well-graded soils generally have fairly high maximum densities at low optimum moisture contents, while clay soils have lower densities. The edge-to-side bonds between clay particles resist compactive efforts to force them into a denser structure. With granular soils, the more well-graded soils have spaces between large particles that fill with smaller particles when compacted, leading to a higher density than with uniform soils. Some typical moisture-density curves are shown in fig. 3–4. Note that a line joining the peak points of the density curves would be approximately parallel to the ZAV curve. This is due to the fact that most soils at their maximum density still contain about 2%–3% air.

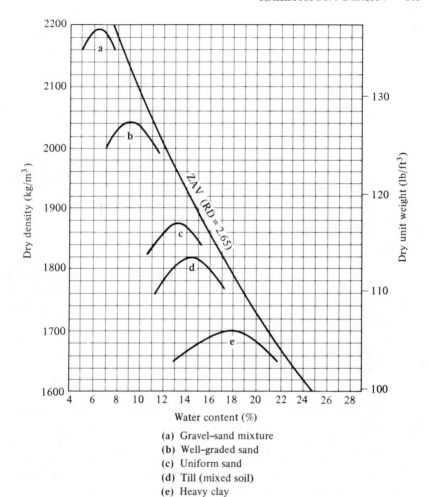

(a) Gravel–sand mixture
(b) Well–graded sand
(c) Uniform sand
(d) Till (mixed soil)
(e) Heavy clay

FIG. 3–4. Typical compaction curves.

3–1.6 Because compaction equipment has become much more effective since Proctor's time, and since the loads imposed on pavements—notably by airplanes —have increased tremendously, a revised test using a much higher compactive effort is now often used. Called the *modified compaction test* (*modified moisture–density* or *modified Proctor*), the compactive effort is provided by a 4.5-kg (10-lb) hammer, falling 45 cm (18 in) on each of five soil layers in the same mold. The maximum dry density obtained in the modified test is naturally higher than that obtained in the standard test, and occurs at a lower optimum moisture content.

3–1.7 Since the compactibility of soils varies considerably, the construction requirements for roads, dams, and so forth are usually specified as a percentage of the maximum dry density found in a laboratory compaction test for each soil type encountered on the project. For example, a project specification might require that the soil be compacted to 95% of the maximum dry density found by the standard compaction test. During construction, standard compaction tests would be run on each different soil type. If the maximum dry density from the test was 2000 kg/m^3 at an optimum water content of 11%, the required field density would be 95% of 2000, or 1900 kg/m^3. The moisture content of the soil should be as close as possible to 11%, which reduces the required compactive effort (for example, number of passes of the roller).

3–2 FIELD DENSITY

To check compaction on a construction project, field density measurements must be made and compared to the maximum dry density obtained in the laboratory for the soil being compacted. Two main types of field density tests are currently in widespread use:

1. Sampling method.
2. Nuclear densometer.

3–2.1 In the *sampling method,* a sample of compacted material is dug out of a test hole in the soil layer being checked. The dry mass of the soil removed and the volume of the hole are measured. The field dry density equals the dry mass divided by the volume originally occupied by the sample. Two basic methods are used to measure the volume of the hole: a balloon filled with a liquid and a sand-cone apparatus.

3–2.2 Using the *balloon apparatus,* the volume of the sample hole is found by forcing a liquid-filled balloon into the test hole. The rubber membrane allows the fluid to fill all the cavities in the test hole. The volume of fluid required to do this is read on a scale on the apparatus. A typical balloon-type densometer is illustrated in fig. 3–5.

Example 3–3:

Mass of the same from the test hole	450.6 g
Dry mass	404.9 g
Volume of the test hole	203.8 cm^3

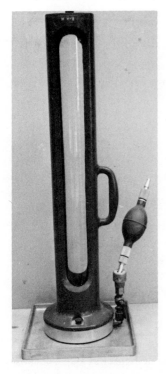

FIG. 3—5. Balloon densometer—the balloon is attached to the bottom and confines the water when it is forced into the hole by the air pump.

Field dry density:

$$\frac{404.9 \text{ g}}{203.8 \text{ cm}^3} = 1.987 \text{ g/cm}^3 = 1987 \text{ kg/m}^3 \ (124.0 \text{ lb/ft}^3)$$

Water content:

$$\frac{(450.6 - 404.9)}{404.9} = 11.3\%$$

3—2.3 With the *sand cone apparatus* for field density determination (see fig. 3—6), the volume of the test hole is obtained from the mass of loose sand required to fill the hole. The sand used is a uniform, medium sand that has an essentially constant loose density when poured into a container or a test hole. The cone acts as a pouring funnel. The mass of a sand container with an attached cone is obtained. The container is then inverted over a test hole, and a valve at the small end of the cone is opened to allow the sand to flow into the test hole.

FIG. 3–6. Sand cone apparatus.

When the flow stops, the valve is closed, and the mass of the sand container with its cone is again measured. The difference between these masses is the mass required to fill the test hole and the cone. As the amount required to fill the cone is constant, the mass in the test hole can be calculated, as can the volume of the test hole, since the loose density of the sand is known.

Example 3–4 – Sand Cone Apparatus: A sand cone holds 851.0 g. The loose density of the sand is 1.430 g/cm^3.

Field Test Results:

Total weight of the soil	639.5 g
Dry weight of the soil	547.9 g
Initial weight of the sand cone apparatus	4527.8 g
Final weight of the sand cone apparatus	3223.9 g

Calculations:

Mass of sand used	4527.8 g – 3223.9 g = 1303.9 g
Mass in test hole	1303.9 g – 851.0 g = 452.9 g
Volume of test hole	$\dfrac{452.9 \text{ g}}{1.430 \text{ g/cm}^3} = 316.7 \text{ cm}^3$
Field dry density	$\dfrac{547.9 \text{ g}}{316.7 \text{ cm}^3} = 1.730 \text{ g/cm}^3 \ (108.0 \text{ lb/ft}^3)$
Field water content	$\dfrac{639.5 - 547.9}{547.9} = 16.7\%$

3–2.4 Quality control of compaction requires that the project meet the specified compaction percentage.

Example 3–5: A dam project specifies 97% compaction. Lab tests on the soil being used indicate that it has a maximum dry density of 1900 kg/m^3 (118.6 lb/ft^3) at an optimum water content of 13.2%. Field tests give the following results: Field dry density is 1790 kg/m^3 (111.7 lb/ft^3) at 16.5% water content.

Results:

1. Compaction percentage = 1790/1900 = 94.2%.
2. Compaction is below requirement.
3. Soil should be allowed to dry out, since field water content is considerably above the optimum.

3–2.5 Sampling methods for determination of field density require a considerable amount of labor, and the results are not readily available. After the sample is taken, its water content must be obtained, as must the volume of the test hole. At least 30 minutes is required (sometimes 6-8 hours if the sample is dried in the laboratory) before test results are available.

The *nuclear method* is becoming more popular, even though the equipment is much more expensive. The test takes only one minute after the surface is prepared, and the results are available immediately. Speed is an important consideration when construction equipment is awaiting results before proceeding.

In a typical nuclear moisture-density tester (fig. 3–7), gamma rays are emitted by a source in the bottom of an index rod. Some of these rays (or photons) are absorbed, and some reach the detector in the front of the gauge. The density of the soil or pavement material can be determined from the number of rays counted by the detector. In the *backscatter position,* the index rod is lowered to the first notch. This opens the shield and allows the source to be exposed in the base of the gauge. Rays are emitted from this source and reflected back to the detector. In the *direct transmission position,* the index rod is lowered into the soil through a prepared hole. The rays are then counted by the detector after passing through the soil. This second position is the more accurate.

Alpha particles are emitted from another source in the base of the gauge. These particles result in neutrons, which lose velocity when they strike the hydrogen atoms in the water molecules. In this way the amount of water in the soil can be measured.

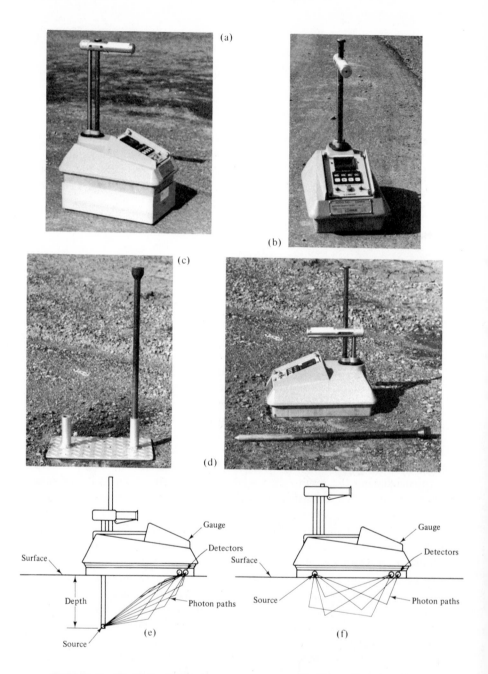

FIG. 3–7. Nuclear densometer—Troxler Model 3401: (a) densometer located on the reference standard; (b) backscatter position for a density test on asphalt; (c) drill rod and guide plate to open hole for; (d) a direct transmission test; (e) direct transmission geometry; (f) backscatter geometry. (Courtesy of Troxler Electronics Laboratories, Inc.)

Example 3–6: The density and moisture standards for a certain nuclear densometer are 3130 and 660, respectively. [Each tester is calibrated periodically (a 5-minute operation in the lab) since radioactivity decreases with age.]

At a test site, the following readings are recorded with the apparatus in the backscatter position:

Moisture count   111
Density count   867

For moisture, the count ratio is 111/660 = 0.168. For density, the count ratio is 867/3130 = 0.277. Therefore, reading from the table included with the test description in section 3–4.3:

and
$$\text{Moisture density} = 124 \text{ kg/m}^3$$
$$\text{Total density} = 2155 \text{ kg/m}^3$$

So dry density is 2155 – 124 = 2031 kg/m^3 and water content is 124/2031 = 6.1%.

3–2.6 In analyzing the results of field compaction tests it may be necessary to make some allowance for the amount of coarse-sized (gravel) particles in the test sample. The laboratory compaction test is usually made with material passing the No. 4 sieve only. If the soil in the field contains a significant amount of gravel particles, the expected density should be revised upwards.

A corrected maximum dry density can be calculated assuming that the gravel particles in a soil composed mainly of finer grains can be compacted to 90% of their theoretical maximum density. For gravel with an RD (G_S) value of 2.65 (often assumed), this would be 90% of 2.65 × 1000 kg/m^3, or 2385 kg/m^3 (90% of 2.65 × 62.4 lb/ft^3, or 148.8 lb/ft^3).

Example 3-7: Laboratory maximum density of a soil is 1900 kg/m^3 (118.6 lb/ft^3). Specifications require 95% compaction. In the field, dry density of the soil is found to be 1810 kg/m^3 (112.9 lb/ft^3). A visual check of the soil in the field indicates that it contains about 20% gravel sizes. (Scales can be used for a more accurate determination of the percentage of gravel.) Check for compaction. Corrected maximum dry density is

$$0.80 \times 1900 + 0.20 \times (90\% \text{ of } 2.65 \times 1000) = 1520 + 477 = 1997 \text{ kg/m}^3$$

$$(0.80 \times 118.6 + 0.20 \times (90\% \text{ of } 2.65 \times 62.4) = 94.9 + 29.8 = 124.7 \text{ lb/ft}^3)$$

Percent compaction is 1810/1997 (112.9/124.7) = 90.6% and is not acceptable.

3–3 COMPACTION METHODS

Soils are compacted in the field by various types of rollers or by mechanical tampers (for trenches, backfill around walls, etc.).

3–3.1 The most common types of rollers are:

1. *For clays and silt-clay mixtures*—sheepsfoot or other types of footed rollers that punch cohesive soils.
2. *For granular soils*—steel-wheeled rollers or rubber-tired rollers.

Rollers are available in a wide range of sizes and masses. Most are also vibratory. Vibrating compactors are especially effective in granular soils.

Types of compaction equipment are illustrated in fig. 3–8.

3–4 TEST PROCEDURES

Procedures for the following tests are presented. (For detailed methods of testing, refer to the standards listed in the Appendix. Refer to the manufacturer's manual for the correct operating instructions for a nuclear densometer.)

3–4.1 Standard Compaction Test
3–4.2 Field Density Using the Sand Cone Apparatus
3–4.3 Density and Water Content with a Nuclear Densometer

3–4.1 Standard Compaction Test

Purpose: To obtain maximum dry density and optimum water content for a soil, using the standard compactive effort.

Theory: This test, also known as the moisture-density test or Proctor test, measures (1) the maximum dry density that a soil can reach under a specified compactive effort, and (2) the moisture content at which maximum dry density is reached. As water is added to a dry soil, the density to which it can be compacted increases because of the lubricating effect of the water. The air in the soil is reduced as the density increases, up to a point of maximum dry density. At this point air content cannot be reduced further, and additional water results in lower density since the excess water must come between soil grains. A plot of dry density vs. water content gives a moisture-density curve. The highest point on this curve is the maximum dry density for this soil at the specified compactive effort. The zero air voids curve gives maximum theoretical densities (no air content) at indicated water contents.

The standard compactive effort is 25 blows of a 2.5-kg (5.5-lb) hammer falling 30 cm (12 in) on each of three layers of material in a mold 10 cm (4 in) in diameter and 943.9 cm^3 (1/30 ft^3) in volume.

Apparatus: standard compaction mold
    standard compaction hammer
    balance
    oven

Procedure:

1. Obtain a sample and break it down to smaller than 4.75 mm (No. 4 sieve).
2. Add water to bring the sample to within about 5% of the estimated optimum water content. Mix thoroughly.
3. Find the mass of the mold without the collar.
4. Compact the soil in the mold (with the collar) in three layers, using standard compactive effort.
5. Remove the collar and trim the compacted mixture even with the top of mold with the straightedge.
6. Measure the mass of the mold and the soil.
7. Extract the soil from the mold, split it, and obtain a sample to test for water content.
8. Place this sample in a beaker, obtain the mass of the sample with the container, and place it in the oven to dry.
9. Break the soil down until it passes a 4.75-mm (No. 4) sieve.
10. Add water to increase the water content by about 2%, and repeat steps 4 through 10.
11. Continue until the mass of the mold and the soil decrease from the previous trial.

Calculations: (see data sheet page 120)

1. Calculate the water content for each trial (w).
2. Calculate the density for each trail (mass of soil/volume of mold).
3. Calculate the dry density for each trial [density/(1 + w)].

Results: Draw the compaction curve on squared paper.
Draw the zero air voids curve with RD (G_S) = 2.65 (if not known).

(a)

(b)

FIG. 3–8. Compaction equipment for soils and granular materials:
(a) a padded wheel vibratory compactor for cohesive soils; (b)
compactor with four padded wheels and a levelling blade; (c) smooth-
wheeled vibratory compactor for granular soils and materials (courtesy of Koehring-Bomag); (d) small vibratory compactor for trenches;
(e) tractor-mounted vibratory plate compactor for trenches.

(c)

(e)

(d)

COMPACTION TEST

| Lab no. _____ Sample no._____ Max. dry density _____ kg/m³ |
| Date _____ Type – Std. – Mod. – Optimum water content _____ (%) |

	Trial no.								
Density	Mass soil + mold	g							
	Mass mold	g							
	Mass soil	g							
	Density	kg/m³							
	Dry density	kg/m³							
Water content	Container no.								
	Mass sample + container	g							
	Mass dry sample + container	g							
	Mass water	g							
	Mass container	g							
	Mass dry soil	g							
	Water content	%							

3–4.2 Field Density Using the Sand Cone Apparatus

Purpose: To obtain the dry density of a compacted soil.

Theory: A sample of soil is dug out of a compacted soil layer, and its dry mass is measured. The volume that the soil occupied is measured by filling the test hole with a uniform sand that has a fairly constant loose dry density. Knowing the sand density and the mass of the sand in the hole, the volume of the test hole can be calculated. Knowing the hole's volume and the dry mass of soil from the hole, the dry density of the soil can be calculated.

In pouring the sand into the test hole, the sand cone acts as a funnel; the base plate acts as the flat, firm surface upon which the cone rests. Therefore, to find the quantity in the test hole, the amount of sand in the cone (and plate) must be subtracted from the mass used. The sand cone apparatus is calibrated periodically to find the loose density of the sand and the mass required to fill the cone and plate.

Note: This test can be done in the laboratory on samples of soil compacted in pans.

Apparatus: sand cone apparatus
              oven

Procedure:

1. Measure the mass of the sand, container, and cone.

2. Place a plate on the surface, levelling it so that the soil surface is flat and the plate is in tight contact with the soil.

3. Dig out the soil about 100 mm (4 in) deep, with the same diameter as the hole in the plate. Clean all loose soil out of the test hole with a brush and spoon. Place it in a container and seal to prevent evaporation.

4. Invert the sand container over the hole. Open the valve.

5. When the sand stops flowing, close the valve, remove the container from the test hole, and obtain the mass of the sand, container, and cone.

6. Obtain the total mass and the dry mass of the soil from the test hole.

Results:

Calibration data for sand cone
   Loose density   _____ g/cm^3 (ρ_{sand})
   Mass in cone   _____ g (MC)
Mass of sand + container + cone (initial)   _____ g (A)
                              (final)   _____ g (B)
Mass of soil from test hole   _____ g M
Dry mass of soil   _____ g M_D

Calculations:

Mass of sand used [(A) – (B)]   _____ g (E)
Mass of sand in cone (MC)   _____ g
Mass of sand in test hole [(E) - (MC)]   _____ g (F)
Volume of test hole [(F)/ρ_{sand}]   _____ cm^3 V
Dry density of soil (M_D/V)   _____ g/cm^3
Water content of soil (($M - M_D$)/M_D)   _____ %

3–4.3 Density and Water Content with a Nuclear Densometer

[*Note:* These instructions are intended to give a general indication of the procedure for using this type of apparatus, but specifically apply only to the Troxler Model 3401. To operate any nuclear apparatus, you must refer to the manufacturer's manual. Only authorized operators can use such a meter.]

Purpose: To measure density and water content of soils and asphalt pavements.

Theory: Density and moisture-density in kg/m^3 or lb/ft^3 are related to the number of radioactive rays counted by the detector located at the front of the gauge. This count, divided by a standard count, give the *count ratio* from which the density values are obtained from a table (see table 3–1). A set of tables for each particular meter is included with that meter.

The standard count is made each day, as follows:

1. Place the reference standard (supplied with the meter) on a dry, flat, solid surface well away from any walls and at least 10 m from any other meter. (Truck beds and floors over open spaces are not suitable.)
2. Turn on the power.
3. Place the densometer on the reference standard with the front frame of the meter pulled tight to the raised edge of the standard.
4. After the switch has been on for 10 minutes, turn the time switch to SLOW and press STD button.
5. After four minutes, the ERR symbol will disappear. Press MS and record the moisture standard. Press DS and record the density standard.

The meter can be operated in either the backscatter position or the direct transmission position at depths of 5, 10, 15, and 20 cm (2, 4, 6, and 8 in) below the surface. The backscatter position is suitable for asphalt paving or soils. For a more accurate determination of soil density, use the direct transmission position. Three time periods are available for tests: FAST (0.25 min) is suitable for asphalt paving or other tests where water content is not important; NORM (1 min) is used for routine soils testing; and SLOW (4 min) is required for the daily calibration and special tests.

Apparatus: nuclear densometer

Procedure:

A. *Surface test (Backscatter position)*
   1. Flatten the surface of the soil with a scraper blade, removing all loose material. If the scraping action dislodges any surface stones or lumps, remove them, fill the voids with fine material, and tamp the surface lightly.
   2. Place the densometer on the prepared base. Lower the index rod to the backscatter position.
   3. Set the time on NORM; depress M&D. (For asphalt paving or other material where moisture is not important, time could be set for FAST.)
   4. After one minute the error light will go out. Press MC and read the moisture count. Press DC and read the density count.

B. *Direct transmission position*

1. Scrape the surface smooth, as in #1 above.
2. Place the scraper plate on the surface. Drive the drill rod into the soil at least 5 cm (2 in) deeper than the depth at which the instrument is to be used.
3. Remove the rod. Mark the edge of the scraper plate.
4. Place the gauge on the surface. Lower the source rod to the desired depth.
5. Pull the instrument forward so that rod is in contact with the soil at the front of the hole.
6. Place time on NORM. Depress M&D.
7. After one minute the error light will go out. Press MC and read moisture count. Press DC and read density count.

Table 3–1

MOISTURE DENSITY (kg/m^3)

Count Ratio*	Moisture Density*
0.122	84
0.126	88
0.131	92
0.136	96
0.140	100
0.145	104
0.149	108
0.154	112
0.159	116
0.163	120
0.168	124
0.172	128
0.177	132
0.181	136
0.186	140
0.191	144
0.195	148
0.200	152
0.204	156
0.209	160

* Typical count-ratio–density relationships for the Troxler Model 3401 Nuclear Gauge.

Results: Record the density count (DC) and the moisture count (MC).

Calculations: Divide DC by DS to obtain the count ratio. Find the total density in (kg/m^3) from table 3–2.
    Divide MC by MS to obtain the count ratio for moisture. Find the moisture density (in kg/m^3) from table 3–1.

Dry density   = total density – moisture density
Moisture content = moisture density/dry density

Note: Assume for calculations that MC = 660 and DC = 3130, and use table 3–2.

Table 3–2
TOTAL DENSITY (kg/m^3)

BACKSCATTER POSITION		DIRECT TRANSMISSION POSITION AT 10 cm (4 in) DEPTH	
Count Ratio*	Total Density*	Count Ratio*	Total Density*
0.301	2067	1.669	1595
0.299	2075	1.653	1603
0.297	2083	1.638	1611
0.295	2091	1.623	1619
0.292	2099	1.608	1627
0.290	2107	1.593	1635
0.288	2115	1.578	1643
0.286	2123	1.564	1651
0.284	2131	1.549	1659
0.281	2139	1.535	1667
0.279	2147	1.521	1675
0.277	2155	1.507	1683
0.275	2163	1.493	1691
0.273	2171	1.479	1699
0.271	2179	1.465	1707
0.269	2188	1.452	1715
0.267	2196	1.438	1723
0.265	2204	1.425	1731
0.263	2212	1.412	1739
0.261	2220	1.398	1747

* Typical count-ratio–density relationships for the Troxler Model 3401 Nuclear Gauge.

3–5 PROBLEMS

Note: Problems 3–1 to 3–6 are general, 3–7 to 3–13 are in SI units and 3–14 to the end are in traditional units.

3–1. Of what use is the ZAV curve?

3–2. Compaction requirements for a highway are not specified in terms of kg/m^3 or lb/ft^3. Why?

3–3. Describe the methods of finding the volume of a test hole in sampling methods used to measure field density.

3–4. What is the ratio of the compactive effort in the modified compaction test to the effort in the standard test?

3–5. What difference in maximum dry densities and optimum water contents would you expect between sands and clays? Why?

3–6. Using the nuclear densometer described in the text, the density count obtained is 924 (backscatter position) and the moisture count is 135. Find the dry density and the moisture content.

3–7. Results of a standard compaction test are:

Trial No. 1	Mass of soil + mold		5619	g
	Mass of mold		3735	g
	Water content test			
	Mass of sample + container		288.26 g	
	Mass of dry sample + container		265.39 g	
	Mass of container		104.31 g	
Trial No. 2	Dry density	$1795 \ kg/m^3$,	$w = 14.9\%$	
Trial No. 3	Dry density	$1828 \ kg/m^3$,	$w = 15.6\%$	
Trial No. 4	Dry density	$1832 \ kg/m^3$,	$w = 16.5\%$	
Tiral No. 5	Dry density	$1796 \ kg/m^3$,	$w = 17.6\%$	

Complete the calculations for Trial No. 1. Draw the moisture-density curve. Include ZAV curves at $w = 16\%$, 17%, and 18%. (The relative density of the soil is 2.65.) Find the maximum dry density and the optimum water content.

3–8. In a field density test the following results were recorded:

Volume of test hole		$865.9 \ cm^3$
Mass of soil from test hole		1923 g
Moisture content sample of this soil	(original mass)	147.3 g
	(final mass)	135.0 g

Find the field dry density to be used in checking for compaction.

3–9. Find ZAV total and dry densities for a soil with RD = 2.68 at (a) w = 0%, and (b) w = 15%.

3–10. A compaction test on a sand soil resulted in three points only for the compaction curve as follows:

Trial 1	1	2	3
Water content (%)	7.0	7.5	8.5
Dry density (kg/m^3)	2124	2160	2144

RD (G_S) for the soil is 2.65. Plot the ZAV curve at w = 7, 8 and 9%, and the test points. Draw the compaction curve using the fact that the curve is parallel to the ZAV line on the wet side of optimum. Find maximum dry density and optimum water content.

3–11. Specifications for a highway require that the soil be compacted to 95% of standard laboratory dry density. Tests on a soil in one section of the road indicate that it has a maximum dry density of 1950 kg/m^3 at an optimum water content of 11.8%. Field density tests are conducted at five locations. Report on the results detailed below. Is compaction satisfactory? Should water be added, or should the road be allowed to dry? Why?

(a) Nuclear densometer:

Total density	2090 kg/m^3
Water content	11.0%

(b) Test hole:

Volume	917.7 cm^3
Total mass of soil	2046 g
Dry mass of soil	1822 g

(c) Test hole:

Volume	1003 cm^3
Total mass of soil	1986.3 g

Sample of soil tested for moisture content:

| Original mass | 199.5 g |
| Dried mass | 183.7 g |

(d) Test hole:

Volume	820.5 cm^3
Total mass of soil	1668.0 g
Water content of soil	14.5%

(e) Nuclear densometer:

| Total density | 1985 kg/m^3 |
| Moisture density | 172 kg/m^3 |

3—12. (a) Specifications for a dam project require compaction equal to 98% of laboratory maximum dry density.

(b) Proctor compaction tests on the soil being used result in a maximum dry density of 2106 kg/m^3 and an optimum water content of 8.2%.

(c) The sand cone test apparatus is calibrated in the laboratory, as follows:

| Mass required to fill the cone | 773 g |
| Mass required to fill a Proctor mold (to find ρ_{sand}) | 1421 g |

(d) The field density test is conducted, with following results:

Mass of sand, bottle, and cone before test	6491 g
Mass of sand, bottle, and cone after test	3217 g
Mass of sample from test hole plus container	3820.5 g
Mass of container	123.5 g
Water content test on sample from test hole:	
Total mass plus container	283.12 g
Mass of container	91.33 g
Dry mass plus container	263.82 g

What results would you report?

3—13. Specifications require 96% compaction on a project. Maximum dry density was found to be 1830 kg/m^3 at a water content of 13.5% in laboratory tests. Results of field density test were dry density = 1780 kg/m³ and water content 11.0%. Sample in the field contained 15% gravel sizes. Is the compaction satisfactory?

3—14. Results of a standard compaction test are:

Trial No. 1	Mass of soil + mold	12.39 lb
	Mass of mold	8.23 lb
	Water content test	
	Mass of sample + container	288.26 g
	Mass of dry sample + container	265.39 g
	Mass of container	104.31 g
Trial No. 2	Dry density 112.0 lb/ft³,	w = 14.9%
Trial No. 3	Dry density 114.1 lb/ft³,	w = 15.6%
Trial No. 4	Dry density 114.3 lb/ft³,	w = 16.5%
Trial No. 5	Dry density 112.0 lb/ft³,	w = 17.6%

Complete the calculations for Trial No. 1. Draw the moisture-density curve. Include ZAV curves at w = 16%, 17%, and 18%. (The specific gravity of the soil is 2.65.) Find the maximum dry unit weight and the optimum water content.

3—15. In a field density test the following results were recorded:

Volume of test hole		0.03058 ft³
Mass of soil from test hole		4.239 lb
Moisture content sample of this soil	(original mass)	147.3 g
	(final mass)	135.0 g

Find field dry unit weight (density).

3—16. Find ZAV total and dry unit weights for a soil with G_S = 2.68 at (a) w = 0%, and (b) w = 15%.

3—17. A compaction test on a sand soil resulted in three points only for the compaction curve as follows:

Trial 1	1	2	3
Water content (%)	7.0	7.5	8.5
Dry unit weight (lb/ft³)	132.5	134.8	133.8

G_S for the soil is 2.65. Plot the ZAV curve at w = 7%, 8%, and 9%, and the test points. Draw the compaction curve using the fact that the curve is parallel to the ZAV line on the wet side of optimum. Find maximum dry unit weight and optimum water content.

3—18. Specifications for a highway require that the soil be compacted to 95% of standard laboratory dry density (unit weight). Tests on a soil in one section of the road indicate that it has a maximum dry density of 121.7 lb/ft^3 at an optimum water content of 11.8%. Field density tests are conducted at five locations. Report on the results detailed below. Is compaction satisfactory? Should water be added, or should the road be allowed to dry? Why?

(a) Nuclear densometer:

Total density	130.4 lb/ft^3
Water content	11.0%

(b) Test hole:

Volume	0.03241 ft^3
Total mass of soil	4.511 lb
Dry mass of soil	4.018 lb

(c) Test hole:

Volume	0.03542 ft^3
Total mass of soil	4.379 lb

Sample of soil tested for moisture content:

Original mass	199.5 g
Dried mass	183.7 g

(d) Test hole:

Volume	0.02896 ft^3
Total mass of soil	3.677 lb
Water content of soil	14.5%

(e) Nuclear densometer:

Total density	123.9 lb/ft^3
Moisture density	10.7 lb/ft^3

3–19. (a) Specifications for a dam project require compaction equal to 98% of laboratory maximum dry unit weight (density).

  (b) Proctor compaction tests on the soil being used result in a maximum dry density of 131.4 lb/ft^3 and an optimum water content of 8.2%.

  (c) The sand-cone test apparatus is calibrated in the laboratory, as follows:

Mass required to fill the cone	773 g
Mass required to fill a Proctor mold (to find ρ_{sand})	1421 g

  (d) The field density test is conducted, with following results:

Mass of sand, bottle, and cone before test	14.310 lb
Mass of sand, bottle, and cone after test	7.092 lb
Mass of sample from test hole plus container	8.423 lb
Mass of container	0.272 lb
Water content test on sample from test hole:	
Total mass plus container	283.12 g
Mass of container	91.33 g
Dry mass plus container	263.82 g

  What results would you report?

3–20. Specifications require 96% compaction on a project. Maximum dry unit weight was found to be 114 lb/ft^3 at a water content of 13.5% in laboratory tests. Results of field unit weight test were dry unit weight = 111 lb/ft^3 and water content 11.0%. Sample in the field contained about 15% gravel sizes. Is the compaction satisfactory?

4

Aggregates

Aggregates are granular mineral particles used either in combination with various types of cementing material to form concretes, or alone as road bases, backfill, etc. Some typical uses of aggregates are portland cement concrete, asphalt concrete, asphalt surfaces, road bases and subbases, railroad ballast, trench backfill, fill under floor slabs, concrete blocks, water filtration beds, drainage structures, riprap, and gabion material.

Properties required in an aggregate depend on its proposed use. But the types of aggregates, their basic properties, and tests used to evaluate these properties apply to most uses. Detailed requirements for various types of construction are discussed in later chapters.

4–1 AGGREGATE SOURCES

Sources of aggregates for construction include:

— natural sand and gravel deposits
— crushed rock
— slag and mine refuse
— rubble and refuse
— artificial aggregates
— pulverized concrete and asphalt pavements

The first two sources supply the bulk of the aggregates used.

4–1.1 Slag is produced as a byproduct of iron production in blast furnaces; it is available in areas close to steel plants. The types that are produced vary considerably. Because of their low density, some slags are used as aggregates in lightweight or insulating concrete. Ground slag is also used as a cementing agent, replacing part of the portland cement in weak concrete. Other types of mine and mill refuse and tailings are used locally as aggregates.

Building rubble, incinerator residue, and other types of refuse such as glass chips have been used as aggregates.

Artificial aggregates include (1) expanded shale, used in lightweight concrete, and (2) other special materials, such as plastics, used for lightweight or insulating construction.

4–1.2 Natural sand and gravel deposits (sand and gravel pits) have been used extensively for aggregates. These consist of sand or gravel soils which have been (1) sorted to eliminate most of the silt and clay sizes, and then (2) deposited in glacial formations (for example, eskers and outwash plains), river deposits or along beaches of lakes and seas.

The material is loose, and is usually excavated with power shovels or front end loaders. Often it is crushed, especially if there are cobbles or boulders in the deposit. The smaller sizes go through the crusher without change, whereas larger particles are broken down to the desired size. *Crushed gravel,* as this is called, is a higher-quality aggregate for many uses.

These aggregates are often processed through a washing plant, which cleans the dust off the particles and also removes the silt and clay particles or alters the gradation of the aggregate in other ways.

A deposit might be composed of many different types of mineral particles —such as limestone, sandstone, and granite—depending on the original rock formation from which the particles came. A typical gravel pit is shown in fig. 4–1.

4–1.3 The properties of aggregates produced in quarries from bedrock basically depend on the type of bedrock. There are three major classes of rock—igneous, sedimentary, and metamorphic.

Igneous rocks were the original rock, formed from the cooling of molten material. Coarse-grained igneous rocks—such as granite—cooled slowly. Fine-grained igneous rocks—such as basalt and trap rock—cooled more quickly.

Sedimentary rocks were formed from the solidification of chemical or mineral sediments deposited under ancient seas. They are usually layered since the original material was deposited in this manner. Some typical sedimentary rocks, with their composition, are:

Limestone	Calcium carbonate
Dolomite	Calcium carbonate and magnesium carbonate
Shale	Clay

Sandstone	Sand
Gypsum	Calcium sulphate
Conglomerate	Gravel
Chert	Fine sand

Metamorphic rocks are igneous or sedimentary rocks that have been changed (metamorphosed) due to intense heat and pressure. Examples are:

Slate	From shale
Marble	From limestone
Quartzite	From sandstone
Gneiss	From granite

Igneous and metamorphic rocks are usually very hard and make excellent aggregates for most purposes.

Limestone and dolomite are quite common sedimentary rocks. They are softer than igneous rocks, but are still acceptable as aggregates for most purposes.

Shale, being composed of clay grains, is very weak and disintegrates easily when exposed to the weather. It is a poor aggregate.

Chert is often found as an impurity in other rock deposits, and may also disintegrate when exposed to weathering.

Aggregates produced from bedrock are obtained from *quarries*. A substantial face of rock (5 m - 20 m or more) is exposed. Holes are drilled from the surface. Then dynamite is placed in these holes to break the rock into sizes that can be transported. The rock is then crushed to the required sizes in various types of rock crushers. Figure 4–2 shows a modern dolomite quarry.

4–2 AGGREGATE TERMS AND TYPES

Aggregates are very common materials; the terms used to describe them are many and varied. These descriptive terms are based on source, size, shape, type, use, and other properties.

4–2.1 Some typical terms used in describing aggregates are:

1. *Fine aggregate* (sand sizes)—aggregate particles mainly between 4.75 mm (No. 4 sieve) and 75 μm (No. 200 sieve) in size.
2. *Coarse aggregate* (gravel sizes)—aggregate particles mainly larger than 4.75 mm (No. 4 sieve).
3. *Pit run*—aggregate from a sand or gravel pit, with no processing.

FIG. 4–1. Gravel pit: (a) pit face, showing layers of sand and gravel (kame deposit); (b) loading with a front end loader; (c) screening plant to divide materials into different sizes; (d) stockpiles of processed aggregates. (Courtesy of D. Cruppi and Sons.)

(a)

(b)

(c)

(d)

FIG. 4–2. Aggregate quarry: (a,b) quarry face; (c,d,e) blasting; (f) transporting to the crusher; (g,h) gyratory crusher breaks rocks to 20 cm (8-in), size; (i) secondary crushers (in background) reduce the size further; (j) aggregates are screened and stockpiled in bins for loading. (Courtesy of Dufferin Materials and Construction.)

135

FIG. 4-2. (Continued.)

4. *Crushed gravel*—pit gravel (or gravel and sand) that has been put through a crusher either to break many of the rounded gravel particles to a smaller size or to produce rougher surfaces.
5. *Crushed rock*—aggregate from the crushing of bedrock. All particles are angular, not rounded as in gravel.
6. *Screenings*—the chips and dust or powder that are produced in the crushing of bedrock for aggregates.
7. *Concrete sand*—sand that has been washed (usually) to remove dust and fines.
8. *Fines*—silt, clay or dust particles smaller than 75 μm (No. 200 sieve), usually undesirable impurities in aggregates.

4–2.2 Aggregate and sieve sizes commonly used in construction are indicated in table 4–1.

Table 4–1

AGGREGATE AND SIEVE SIZES COMMONLY
USED IN CONSTRUCTION

Sieve Designation		Suggested Standard Sizes	
Traditional	*Metric*	*Used by the Concrete Industry in the United States*	*Used by the Concrete Industry in Canada*
3-in	75 mm	70 mm	80 mm
2-in.	50 mm	50 mm	56 mm
1½-in.	38* mm	40 mm	40 mm
1-in.	25.0 mm	25 mm	28 mm
3/4-in.	19.0 mm	20 mm	20 mm
1/2-in.	12.5 mm	12.5 mm	14 mm
3/8-in.	9.5 mm	10 mm	10 mm
No. 4	4.75 mm	5 mm	5 mm
No. 8	2.36 mm	2.5 mm	2.5 mm
No. 16	1.18 mm	1.2 mm	1.25 mm
No. 30	600 μm	600 μm	630 μm
No. 50	300 μm	300 μm	315 μm
No. 100	150 μm	150 μm	160 μm
No. 200	75 μm	75 μm	80 μm

* also designated as 37.5 mm or 38.1 mm

4–2.3 A term often used in describing and specifying aggregates is nominal size. (Size determination by sieves for aggregates is based on the mass retained and passing each sieve as illustrated for soils in section 1–3).

It is not necessary that 100% of the particles of an aggregate be within the specified size range for construction purposes. A small amount, usually 5% or 10%, is allowed to be either larger or smaller than the specified size, as it would be economically impossible to ensure that 100% of the particles are within any specified range. Therefore if 19 mm (3/4 in) is the maximum size of aggregate desired for a concrete mix, specifications will indicate that the nominal maximum size is 19 mm. In this case, 90% of the sample (minimum) must be smaller than 19 mm, and 100%, smaller than the next higher standard size, 25 mm (1 in). (Table 4–1 shows the standard sizes usually specified.) Fine aggregate has a nominal maximum size of 4.75 mm (No. 4 sieve). Therefore specifications will require that 100% of the aggregate pass the 9.5 mm (3/8 in) sieve, and 90 (or 95%) pass 4.75 mm.

For coarse aggregates a nominal size range is often used to specify aggregate materials. A 19-4.75 mm aggregate is a material with 100% passing the 25-mm sieve, a 90%–100% passing the 19-mm sieve, and 0%–10% passing the 4.75 mm sieve.

A single size coarse aggregate is called clear. Most of the particles are between the specified maximum size and a minimum size, which is one-half of the maximum. For example, a 19-mm clear aggregate will have 100% smaller than 25 mm, 90%–100% smaller than 19 mm, and 0%–10% smaller than 9.5 mm.

Figure 4–3 shows six different types of aggregates.

(a)   (b)   (c)

(d)   (e)   (f)

FIG. 4–3. Aggregates: (a) concrete sand, (b) 19 mm (3/4 in) clear crushed rock (limestone); (c) crushed gravel (gravel and sand); (d) screenings; (e) 9.5 mm (3/8 in) clear gravel (pea gravel); (f) 9.5 mm (3/8 in) clear crushed (a hard rock igneous rock called trap rock).

4–3 PROPERTIES

Important properties of aggregates include:

— gradation
— relative density (specific gravity) and absorption
— hardness (resistance to wear)
— durability (resistance to weathering)
— shape and surface texture
— deleterious substances
— crushing strength

4–3.1 Gradation (or *grain size analysis*) is the most common test performed on aggregates.

Most specifications for highway bases, concrete and asphalt mixes require a grain size distribution that will provide a dense, strong mixture. This is accomplished by ensuring that the shape of the grain size distribution curve is similar to those shown in fig. 4–4. This shape assures maximum density and strength. The voids between the larger particles are filled with medium particles. The

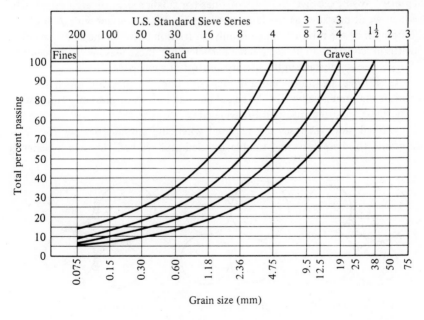

FIG. 4–4. Maximum density curves for aggregates using the Fuller Relationship.

remaining voids are filled with still smaller particles, until the smallest voids are filled with a small amount of fines. Figure 4—4 shows the theoretical maximum density curves for aggregates, based on the Fuller maximum density equation.

Strength, or resistance to shear failure, in road bases and other aggregates that carry loads is increased greatly if the mixture is dense graded. The larger particles are in contact with each other, developing frictional resistance to shearing failure, and tightly bound together due to the interlocking effect of the smaller particles. (A shear failure results when some particles slide over others as in the development of ruts in a roadway.)

Figure 4—5 illustrates the effect of gradation on the strength of aggregate mixtures.

High-density mixtures are also important for the purpose of economy. In concrete and asphalt mixtures, the cementing agent must coat each particle and fill most of the voids between particles to give a strong mixture. If the relatively cheaper portion of the mix—that is, the aggregates—fills most of the voids between large particles, a more economical mix is possible.

Often the fines content must be limited. The silt and clay particles (finer than 75 μm or No. 200 sieve) are relatively weak. Covering them requires an excessive amount of cement. If fines are present as dust on larger particles, they weaken the bond between the cement and those particles. Fines in highway bases may lead to drainage and frost-heaving problems, as discussed in more detail in chapter 5. Also excessive amounts of fines (or smaller sizes of aggregates) may result in weak mixtures as the large particles are not in contact with each other. The strength of the mixture depends only on friction between the small particles, which is much less than between large particles. For these reasons the percentage of fines is very important in the quality control of aggregates.

Clay fines are more harmful than silt fines as they are much smaller. Therefore maximum values for the liquid limit and index of plasticity (indicators of clay fines) are often specified for aggregates.

Washed sieve analyses are required when the amount passing 75 μm is important. In these tests the sample is dried and washed. The wash water is

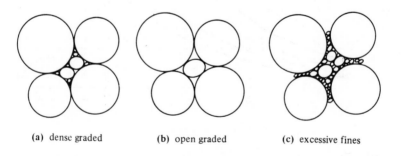

(a) dense graded   (b) open graded   (c) excessive fines

FIG. 4—5. Aggregate gradations.

poured out over a 75-μm sieve, and the material retained in the sieve is returned to the sample. It is then dried again and dry-sieved. The total amount passing 75 μm is the sum of the amounts lost in washing and passing the 75-μm sieve.

Example 4–1:

Mass of sample	446.7 g
Mass after washing	414.1 g

Results of dry sieving:

Retained in 4.75 mm	0.0 g
1.18 mm	205.3 g
300 μm	127.9 g
75 μm	76.4 g
Pan	3.8 g

Find the grain size distribution:

Lost in washing over 75 μm	32.6 g
Passing 75 μm in sieving	3.8 g
Total finer than 75 μm	36.4 g

Calculations:

Sieve	Retained	Percentage Retained	Cumulative Passing
4.75 mm	0.0 g	0.0	100 %
1.18 mm	205.3	46.0	54.0
300 μm	127.9	28.7	25.3
75 μm	76.4	17.1	8.2
Pan	36.4	8.2	
	446.0 g	100.0%	

(*Note*:   Only 0.7 g is lost during sieving, which is an acceptable loss.)

4–3.2 The *relative density* (specific gravity) and *absorption* of aggregates are important properties, especially in mix design for concrete and asphalt mixtures.

In mix design, it is necessary to measure accurately the volumes occupied by specified masses of aggregate. Therefore the voids in the aggregate must be considered.

Aggregates are porous materials, and water can seep into the pores of the particles. The relative density is calculated from:

$$RD = \frac{M}{V \times \rho_W}$$

For aggregates, the mass can be either the dry mass or the total mass (including absorbed water). The volume can be either the bulk volume or the net volume (the bulk volume minus the volume of absorbed water). Aggregates can also have surface moisture, which is moisture on the surface in addition to the absorbed water. These moisture conditions are illustrated in fig. 4–6.

Relative density calculations are made as follows:

Apparent	$RD_A = M_D/(V_N \times \rho_W)$	*(4–1)*
Bulk	$RD_B = M_D/(V_B \times \rho_W)$	*(4–2)*
Saturated, surface-dry	$RD_{SSD} = M_{SSD}/(V_B \times \rho_W)$	*(4–3)*
Percentage absorption	$\% \text{ Abs} = M_{WA}/M_D$	*(4–4)*

(a) *Dry* — mass dry = M_D ; volume net = V_N.

(b) *Saturated, surface–dry* — all permeable pores filled with water; particle appears moist but is not shiny.
Mass saturated, surface-dry = M_{SSD} (includes mass of particle and mass of absorbed water, M_{WA} ($M_{SSD} = M_D + M_{WA}$);
volume bulk = V_B (includes volume of absorbed water).

(c) *Wet* — pores saturated and surface covered with a film of water; particle is shiny.

FIG. 4–6. Moisture conditions in aggregates.

(Using traditional units, G_A, G_B, and G_{SSD} represent the specific gravity values in these conditions. Calculations are the same.)

Example 4-2: The dry mass of a sample is 2239.1 g. The mass in saturated, surface-dry condition is 2268.4 g. The net volume is 835.4 cm³. Find the relative density values.

Mass of absorbed water	$2268.4 - 2239.1 = 29.3$ g
Volume of absorbed water	$\dfrac{29.3 \text{ g}}{1 \text{ g/cm}^3} = 29.3 \text{ cm}^3$
Bulk volume	$835.4 + 29.3 = 864.7 \text{ cm}^3$

Therefore

$$RD_A = \frac{2239.1 \text{ g}}{835.4 \text{ cm}^3 \times 1 \text{ g/cm}^3} = 2.68$$

$$RD_B = \frac{2239.1 \text{ g}}{864.7 \text{ cm}^3 \times 1 \text{ g/cm}^3} = 2.59$$

$$RD_{SSD} = \frac{2268.4 \text{ g}}{864.7 \text{ cm}^3 \times 1 \text{ g/cm}^3} = 2.62$$

$$\% \text{ Abs} = \frac{29.3 \text{ g}}{2239.1 \text{ g}} = 1.31\%$$

Example 4-3: Given that the bulk relative density (RD_B) is 2.62 and that absorption is 1.2%, find the relative density in saturated, surface-dry condition. To simplify the solution, a mass or volume can be assumed. As the RD_B (M_D/V_B) is known, assume $V_B = 1000$ cm. Therefore

$$M_D = 2620 \text{ g}$$

$$M_{WA} = 0.012 \times 2620 = 31 \text{ g}$$

$$M_{SSD} = 2620 + 31 = 2651 \text{ g}$$

$$RD_{SSD} = \frac{2651 \text{ g}}{1000 \text{ cm}^3 \times 1 \text{ g/cm}^3} = 2.65$$

4-3.3 Hardness (or resistance to wear) is important in many applications. It is important that the aggregates for pavement surfaces do not become rounded or polished due to traffic, since they would then become less skid-resistant. Floors subject to heavy traffic in factories must be wear-resistant. Aggregates in road-beds are subjected to many applications of loads, and should not be allowed to wear down due to movement between the particles.

The *Los Angeles abrasion test* (see fig. 4—7) is a commonly accepted measure of the hardness of aggregates. Aggregates are prepared and placed in a drum with a number of steel balls. The drum is rotated a specified number of times, and the loss of aggregates or the amount ground down is measured.

Example 4—4:

Mass placed in abrasion machine	5015 g
Mass of intact particles left after test	3891 g

$$\% \text{ loss} = \frac{5015 \text{ g} - 3891 \text{ g}}{5015 \text{ g}} = 22.4\%$$

4—3.4 Another important property, especially in areas where frost is common, is the aggregate's *durability* (or resistance to weathering). Aggregate particles

FIG. 4—7. Los Angeles abrasion apparatus.

have pores. Sedimentary rocks sometimes have many planes of weakness between layers. Water enters these pores or openings between layers. On freezing, it expands and opens a wider crack. This allows more water to seep in after thawing. Many cycles of freezing and thawing may break or disintegrate the aggregate particles.

The *soundness test* is commonly used to measure an aggregate's susceptibility to weathering. A sample of aggregate is saturated in a solution of magnesium sulphate or sodium sulphate, and then removed and dried in an oven. This process is repeated for five cycles. On completion, the percentage lost or broken down is calculated.

Example 4–5:

$$\text{Original mass of sample} \quad 2175 \text{ g}$$
$$\text{Mass of particles after test} \quad 1847 \text{ g}$$

$$\% \text{ loss} = \frac{2175 \text{ g} - 1847 \text{ g}}{2175 \text{ g}} = 15.1\%$$

The crystallization of salts in the pores during the drying phase of the test exerts pressure on the sides of the pores and eventually breaks the particles if they are weak. This action is, in theory, similar to the growth of ice crystals in pores filled with water.

Unfortunately the soundness test is not very reliable. Aggregates that pass the test sometimes break up under field conditions, while others that fail the test prove acceptable in field use. However, it is relatively simple to conduct and is still used extensively. A more reliable indication of aggregate quality is given by a visual inspection of the performance of the aggregate in pavements that have been subjected to freezing and thawing for a number of years. Also used are tests that involve saturating samples in water and freezing them for a number of cycles, but these require a great deal more time and equipment.

As has already been mentioned, shale is a very weak rock that disintegrates readily when exposed. Fig. 4–8 shows a sample of shale before and after exposure to the sulphate test.

4–3.5 *Particle shape and surface texture* affect the strength of the aggregate particles, the bond with cementing materials, and the resistance to sliding of one particle over another.

Flat particles, thin particles, or long, needle-shaped particles break more easily than cubical particles. Particles with rough, fractured faces allow a better bond with cements than do rounded, smooth gravel particles.

FIG. 4–8. A sample of shale before and after exposure to soundness test.

Rough faces on the aggregate particles allow a higher friction strength to
be developed if some load would tend to force one particle to slide over an
adjacent particle. Many specifications restrict the percentage of long or thin
particles allowed in aggregates, or require that a percentage of the particles have
at least one fractured face or side that has been broken in a gravel crusher.

4–3.6 *Deleterious substances* are harmful or injurious materials. They include
various types of weak or low-quality particles and coatings that are found on
the surface of aggregate particles.

Deleterious substances include organic coating; dust (material passing the
75-μm sieve); clay lumps; shale; coal particles; friable particles (easy to crumble); chert (which may break up when exposed to freezing and thawing); badly
weathered particles; soft particles; and lightweight particles. These substances
may affect the bond between cements and aggregates or may break up during
mixing or use.

Special tests are often required to measure the amount of specific deleterious substances in aggregates to be used in asphalt mixtures and portland cement
concretes. Evaluation of the results of these tests usually involves finding the
mass of the deleterious substance as a percentage of the test sample.

A general test sometimes used to measure deleterious materials is the
petrographic analysis. Each particle is examined and then rated as excellent,
good, fair, or poor (or rated on some other scale). The amount of particles in
each quality group is summarized. The summary is used to help measure acceptability of the aggregate.

4–3.7 *Crushing strength* is the compressive load that aggregate particles can
carry before breaking. This trait is relatively unimportant for most aggregate
uses since aggregate strength is much greater than the strength of the asphalt or
concrete mixture.

4–3.8 *Chemical stability* refers to specific problems due to the chemical composition of aggregate particles. Aggregates susceptible to stripping may lead to disintegration of asphalt surfaces. Reactive aggregates may cause the break up of portland cement concretes.

These properties are discussed in more detail in subsequent chapters on these materials.

4–3.9   Table 4–2 summarizes this discussion on aggregate quality.

Table 4–2
AGGREGATE QUALITY

Property	Test	Significance
Gradation	Sieve Analysis (washed for fine aggregates)	Strength in base courses and asphalt mixes, economy in concrete.
Fines Content	Washed Sieve Analysis (fine aggregate) Washing Test (coarse aggregate) Atterberg Limits	Strength in base courses and asphalt mixes, drainage and frost problems in highway bases, economy in concrete.
Relative Density (Specific Gravity), and Absorption	Relative Density	Design calculations, mix adjustments.
Hardness	Abrasion	Quality of particle, skid resistance of highway surfaces, wear resistance of floors.
Durability	Soundness	Durability, resistance to weathering.
Particle Shape	Amount of Thin or Elongated Particles	Durability of particle.
Particle Surface	Amount of Crushed Particles	Strength in base courses and asphalt mixes.
Deleterious Particles or Substances	Petrographic Test, Tests for % clay, % coal, etc.	Durability of particles.
Chemical Stability	Reactivity (concrete aggregates), Stripping (asphalt aggregates)	Durability of concrete and asphalt.

4-4 SPECIFICATIONS

For each of the preceding properties of aggregates, specifications vary considerably. Requirements for aggregates to be used as road bases differ from the requirements for aggregates to be used in asphalt or portland cement concretes. Engineering authorities that specify the quality of aggregates also have requirements that differ according to local experience, availability of materials, and type of project.

Specifications suggested by ASTM, AASHTO, and CSA are given in the chapters on pavements, asphalt mixtures, and portland cement concretes. Special properties for aggregates are also discussed in those chapters. Only typical requirements and evaluation techniques are discussed in this chapter.

4-4.1 Typical specifications for a highway base course might be:

1. Gradation test:

Passing 25 mm (1 in.)	100%
19 mm (3/4 in.)	90-100%
9.5 mm (3/8 in.)	50-75%
4.75 mm (No. 4)	35-55%
1.18 mm (No. 16)	15-40%
300 μm (No. 50)	5-22%
75 μm (No. 200)	2-8%

2. Physical property requirements:

Abrasion loss (Los Angeles test)	maximum allowable 40%
Soundness loss (using magnesium sulphate)	maximum allowable 18%

The gradation limits are shown in fig. 4-9.

Example 4-6: The test results on an aggregate to be evaluated as a highway base course are as follows:

1. Gradation test:

Passing 25 mm (1 in.)	100%
19 mm (3/4 in.)	98%

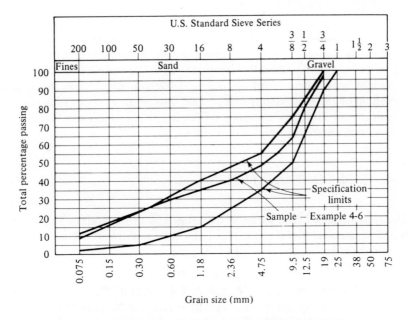

FIG. 4–9. Gradation graph showing specifications limits.

12.5 mm (1/2 in.)	81%
9.5 mm (3/8 in.)	63%
4.75 mm (No. 4)	48%
1.18 mm (No. 16)	35%
300 µm (No. 50)	24%
75 µm (No. 200)	11%

2. Abrasion test:

Original mass	5009 g
Final mass	3267 g

3. Soundness test:

Original mass	2649 g
Final mass	2115 g

Checking for aggregate acceptability according to the specifications above, we find that the gradation test results do not meet the requirements because of an excessive amount of fine-sized material. Note that 24% and 11% pass the 300-μm and 75-μm sieves, respectively, although the maximum allowable percentages are 22% and 8%. The sample meets the abrasion specifications: Loss is 34.8% (1742/5009), below the allowable 40%. The sample does not meet the soundness requirement: Loss is 20.2% (534/2649), greater than the allowable maximum of 18%. The sample fails to meet the specification due to the excessive proportion of fine sizes and poor results in the soundness test. The sample's grain size distribution curve is shown in fig. 4–9.

4–5 SAMPLING AND TESTING

The tests discussed here measure the properties of the aggregate sample being tested. Remember that if this sample is not representative of the aggregates to be used in construction, the tests are of little use.

4–5.1 CSA standard A23.2-1A and ASTM standard D75 give methods to be followed in sampling aggregates in various locations in the field. Sizes of required samples according to ASTM D75 are given in table 4–3.

Table 4–3

SIZE OF SAMPLES (FROM ASTM STANDARD D75)*

Type	Nominal Maximum Size (sieve size)	Approximate Minimum Mass of Field Sample [lb and (kg)]
Fine aggregate	No. 8 (2.36 mm)	25 (10)
	No. 4 (4.75 mm)	25 (10)
Coarse aggregate	3/8-in. (9.5 mm)	25 (10)
	1/2-in. (12.5 mm)	35 (15)
	3/4-in. (19.0 mm)	55 (25)
	1-in. (25.0 mm)	110 (50)
	1½-in. (37.5 mm)	165 (75)
	2-in. (50 mm)	220 (100)
	3-in. (75 mm)	330 (150)

* Reprinted by the permission of the American Society for Testing and Materials, 1916 Race Street, Philadelphia, PA 19103, Copyright.

4–5.2 In conducting laboratory tests, it is very important that the sample tested be representative of the material delivered to the lab. Usually a sample splitter (fig. 4–10) is used to obtain the test sample. The aggregate is split until the required size is obtained.

FIG. 4–10.   Sample splitter.

Table 4–4

SAMPLE SIZE FOR SIEVE ANALYSIS
(MODIFIED FROM ASTM STANDARD C136)*

Type	Nominal Maximum Size (sieve size)	Size of Sample [g (approximate)]
Fine aggregate	No. 8 (2.36 mm)	100
	No. 4 (4.75 mm)	500
		[kg (minimum)]
Coarse aggregate	3/8-in. (9.5 mm)	1
	1/2-in. (12.5 mm)	2
	3/4-in. (19.0 mm)	5
	1-in. (25.0 mm)	10
	1½-in. (37.5 mm)	15
	2-in. (50 mm)	20
	3-in. (75 mm)	60

* Reprinted by the permission of the American Society for Testing and Materials, 1916 Race Street, Philadelphia, PA 19103, Copyright.

4–5.3   The required sample size for each laboratory test is included in the test instructions. Table 4–4 gives the requirements for the most common test—sieve analysis—as specified by ASTM for concrete aggregates.

4–5.4   In testing aggregates composed of significant amounts of both fine and coarse sizes, the sample must be split on the 4.75-mm (No. 4) sieve, and the two fractions sieved separately. Were this not done, the amount of material on the fine sieves would be too large for effective sieving. The sample is first split on the

4.75-mm sieve, and the coarse fraction sieved on coarse sieves down to the 4.75 mm. Material passing 4.75 mm in this operation is added to the fine sample. This fine sample is split down to the required size (about 500 g), washed, dried, and sieved through the fine sieves. To obtain the final grain size distribution curve, the percentage retained on each of these sieves must be multiplied by the ratio of the fine fraction to the whole sample.

Example 4–7: The sample of a granular base material (combined coarse and fine sizes) has an original total mass of 9133 g.
    Sample split on 4.75-mm sieve:

mass retained	5231 g
mass passing	3895 g
mass total	9126 g*

Coarse fraction sieved:

Retained on 19.0 mm	0 g
9.5 mm	2269 g
4.75 mm	2792 g
Pan	167 g
Total	5228 g

Combined fine fraction split down to	496.1 g
After washing over 75 μm and drying, mass is	478.8 g
	17.3 g

Fine sieving:

Retained on 4.75 mm	0 g
1.18 mm	178.3 g
300 μm	215.4 g
75 μm	77.8 g
Pan	6.5 g
Total	478.0 g*

* Totals will usually differ slightly from
  original amounts.

Calculations:

Sieve	Mass Retained	Percentage Retained	Corrected Percentage Retained	Cumulative Percentage Passing
19 mm	0 g	0		100
9.5 mm	2269 g	24.9		75.1
4.75 mm	2792 g	30.6		44.5
Pan	4062 g	44.5		
	(3895 g + 167 g)			
Total	9123 g	100.0%		

Sieve	Mass Retained	Percentage Retained	Corrected Percentage Retained	Cumulative Percentage Passing
4.75 mm	0 g	0		
1.18 mm	178.3 g	36.0	16.0	28.5
300 μm	215.4 g	43.5	19.4	9.1
75 μm	77.8 g	15.7	7.0	2.1
Pan	23.8 g	4.8	2.1	
	(17.3 + 6.5 g)			
Total	495.3 g	100.0%	44.5%	

(Percentage Retained column $\times$ 0.445 = Corrected Percentage Retained)

4–6 BLENDING

To meet the gradation requirements for asphalt or concrete, it is often necessary to blend two or more aggregates together. Charts and diagrams are available to do this blending, but the trial-and-error method is simpler and just about as fast as more complex methods.

4–6.1 Use of the trial-and-error method for blending is illustrated by the following example.

Example 4–8: Three aggregates are to be blended to meet a specification. The aggregates, gradations, and the specification are:

	Aggregate A	Aggregate B	Aggregate C	Specification
Passing 12.5 mm	100%			100%
9.5 mm	62%		100%	72–88%
4.75 mm	8%	100%	78%	45–65%
2.36 mm	2%	91%	52%	30–60%
1.18 mm	0%	73%	36%	25–55%
600 μm		51%	29%	16–40%

300 μm	24%	24%	8-25%
150 μm	4%	20%	4-12%
75 μm	1%	18%	3-6%

Most of the coarse aggregate (larger than 4.75 mm) will come from aggregate A; most of the fines (smaller than 75 μm), from aggregate C. To obtain a mixture that is approximately in the middle of the specification, there should be 55% passing 4.75 mm and 5% passing 75 μm, or 45% larger than 4.75 mm and 5% smaller than 75 μm.

To obtain 45% larger than 4.75 mm, try 45% aggregate A. (This does not all pass 4.75 mm, but aggregate C will add some particles larger than 4.75 mm.)

To obtain 5% smaller than 75 μm, look at aggregate C. For 18% passing 75 μm we would use 100% aggregate C; therefore, for 5% passing 75 μm, we would use 5/18, or 28%, aggregate C. As some smaller than 75 μm are contained in aggregate B, try 25% aggregate C.

Therefore the first trial blend is 45% A, 25% C, and the balance, 30% B. The calculation of the resulting gradation is shown in the table below.

For aggregate A the total used is 45%; therefore,

Passing 12.5 mm $0.45 \times 100\% = 45\%$

Passing 9.5 mm $0.45 \times 62\% = 27.9$

Passing 4.75 mm $0.45 \times 8\% = 3.6$ . . . and so on.

Size	Aggregate A Total Sample	× 45%	Aggregate B Total Sample	× 30%	Aggregate C Total Sample	× 25%	Combined Gradation
Passing 12.5 mm	100%	45.0%	100%	30%	100%	25.0%	100%
9.5 mm	62%	27.9%	100%	30%	100%	25.0%	82.9%
4.75 mm	8%	3.6%	100%	30%	78%	19.5%	53.1%
2.36 mm	2%	0.9%	91%	27.3%	52%	13.0%	41.2%
1.18 mm	0%	0%	73%	21.9%	36%	9.0%	30.9%
600 μm			51%	15.3%	29%	7.2%	22.5%
300 μm			24%	7.2%	24%	6.0%	13.2%
150 μm			4%	1.2%	20%	5.0%	6.2%
75 μm			1%	0.3%	18%	4.5%	4.8%

The combined gradation meets the specifications. If changes were desired, a second trial could quickly be done with changes as indicated by the results of the first trial mix.

4–7   TEST PROCEDURES

Test procedures included in this text are indicated below. As stated previously, you should refer to official standards (listed in the Appendix) for detailed testing methods.

4–7.1   Washed Sieve Analysis (Fine Aggregates)

4–7.2   Relative Density and Absorption (Coarse Aggregates)

4–7.3   Relative Density and Absorption (Fine Aggregates)

4–7.4   Los Angeles Abrasion Test (Coarse Aggregates)

4–7.5   Soundness Test

4–7.6   Gradation Analysis (Combined Aggregates)

4–7.1   Washed Sieve Analysis (Fine Aggregates)

Purpose: To obtain grain size distribution curve for a fine aggregate.

Theory: The sample is dried, placed in a nest of sieves, and shaken. The amount retained on each sieve is weighed; the percentage retained on each sieve and the cumulative percentage passing are calculated. The resulting grain size distribution curve is compared with the specification limits for acceptance. Note the following restrictions:

1. To ensure that the sample is large enough to be representative, a minimum of 400 g is required.

    To ensure that the sample is not too large for effective sieving, a maximum of 600 g is required.

2. To ensure that the percentage passing 75 μm is accurate, the sample is washed over the 75-μm sieve.

3. To ensure that the sample is representative, a sample splitter must be used to obtain test sample.

Apparatus: sieves–9.5 mm, 4.75 mm, 2.36 mm, 1.18 mm, 600 μm, 300 μm, 150 μm, 75 μm
              pan
              sieve shaker
              balance

Procedure:

1. Oven-dry the sample, split it down, and measure the mass.

2. Wash, pouring the wash water out over a 75-μm sieve. Continue until the water is clear. Return the coarse material in the sieve to the sample.
3. Dry. Measure the mass.
4. Place the sample in nest of sieves, then shake.
5. Obtain the mass retained on each sieve.

(*Note:* Total mass passing the 75-μm sieve is the amount washed through 75 μm plus the amount passing 75 μm on dry sieving.)

Results: Calculate the percentage passing each size and plot the grain size distribution curve.

4–7.2 Relative Density (Specific Gravity) and Absorption (Coarse Aggregates)

Purpose: To measure the relative density (apparent; bulk; and saturated, surface-dry) and absorption of a sample of coarse aggregate.

Theory: Aggregates are porous, not solid particles. Water is absorbed by the particle in the pore spaces, which may be relatively shallow or may extend well into the aggregate particle.
    The moisture condition of aggregate particles can be:

1. Dry—oven-dry or no moisture content.
2. Saturated, surface-dry—all pores filled with water, but no moisture film on the surface.
3. Wet—pores saturated and surface moisture present.

    For relative density calculations, either the mass in the dry condition or the mass in the saturated, surface-dry condition can be used. The volume can be the net volume (that is, the volume of the particle, excluding the volume of pore space that can be filled with water) or the bulk volume (that is, the volume of the particle, including pores).
    In this test the particles are soaked, and then their mass is measured (1) in air, (2) submerged, and (3) after drying in the oven. The difference between mass when dry and mass when submerged equals the mass of water displaced by the aggregate. Since the mass of water displaced in grams equals the volume of water displaced in cubic centimeters, the net volume of the aggregate can be obtained.

Apparatus: wire basket
             balance (accurate to 0.1 g)
             oven

Procedure:

1. Wash approximately 2 kg of coarse aggregate. Soak for 24 hours.
2. Pour off the water, then roll the aggregate in a towel until the surface moisture is removed. Wipe the larger pieces individually. The surface moisture
   film, which shines, must be removed, but the particles must not be allowed
   to dry out, as this means that absorbed water is being removed.
3. Obtain the mass.
4. Place the sample in the wire basket and obtain the mass when submerged.
5. Dry the sample in the oven.
6. Measure the mass.

Results:

Mass saturated, surface-dry	_____	g M_{SSD}
Mass submerged	_____	g M_{SUB}
Mass dry	_____	g M_D

Calculations:

Mass of absorbed water $(M_{SSD} - M_D)$	_____	g M_{WA}
Volume net $(M_D - M_{SUB})$	_____	cm^3 V_N
Volume bulk $(V_N + M_{WA})$	_____	cm^3 V_B

Conclusions:

$$RD_A = M_D/V_N \quad = \text{_____}$$
$$RD_B = M_D/V_B \quad = \text{_____}$$
$$RD_{SSD} = M_{SSD}/V_B \quad = \text{_____}$$
$$\text{Absorption} = M_{WA}/M_D = \text{_____}$$

4–7.3 Relative Density and Absorption (Fine Aggregates)

Purpose: To measure the relative density (apparent; bulk; and saturated,
surface-dry) and absorption of a fine aggregate.

Theory: As with coarse aggregates, fine aggregates are porous and absorb water.
Relative density can be calculated using the mass (including or excluding the
mass of absorbed water) and the net or bulk volumes (the latter including the

volume of absorbed water). A sample of wet sand is slowly dried. The moisture film around the sand grains holds the grains together, due to surface tension in the water film. As soon as this surface moisture evaporates, this apparent cohesion between grains disappears. However, at that time the absorbed water, which does not evaporate until the surface water is gone, is still in the aggregate and can, therefore, be measured.

Apparatus: pycnometer (500 ml)
             conical mold and tamper
             balance (accurate to 0.01 g)
             oven

Procedure:

1. Obtain and soak a sample of about 1 kg.
2. Dry the sample slowly with a hair dryer or similar apparatus. While drying, periodically fill the cone with sand, lightly tamp the surface 25 times, and lift the cone to check if the sand maintains the shape of the mold.
3. Continue drying until the sand slumps when the cone is lifted. The sand is then in saturated, surface-dry condition.
4. Place 500.0 g of this sand in the pycnometer. Add water to cover the sand.
5. Roll and agitate the pycnometer to eliminate air bubbles.
6. Adjust the temperature to 23°C (± 2°C) by immersing in water.
7. Fill the pycnometer to the calibrated level.
8. Obtain the total mass.
9. Remove the aggregate from the pycnometer. Dry the sample in oven. Obtain the mass.

Results:

Mass of sand + water + pycnometer	_____	g (C)
Mass of dry sand	_____	g (A)
Mass of pycnometer filled with water at 23°C (usually given)	_____	g (B)

Calculations:

Bulk relative density $RD_B = \dfrac{A}{B + 500 - C} = $ _____

Saturated, surface-dry
relative density
$$RD_{SSD} = \frac{500}{B + 500 - C} = \underline{\hspace{3cm}}$$

Apparent relative density
$$RD_A = \frac{A}{B + A - C} = \underline{\hspace{3cm}}$$

Absorption
$$\% \text{ Abs} = \frac{500 - A}{A} \times 100 = \underline{\hspace{3cm}} \%$$

4–7.4 Los Angeles Abrasion Test (Coarse Aggregates)

Purpose: To measure hardness of aggregates.

Theory: To measure the hardness of aggregates, a sample is placed in a drum with steel balls. The drum is rotated and the balls grind down the aggregate particles. Soft aggregates are quickly ground to dust, whereas hard aggregates lose very little.

Apparatus: Los Angeles abrasion machine
sieves
balance (accurate to 0.01 g)

Sample: Approximately 5000 g of aggregate including 2500 ± 10 g of 19-mm to 12.5-mm (3/4-in. to 1/2-in.) size and 2500 ± 10 g of 12.5-mm to 9.5-mm (1/2-in. to 3/8-in.) size.

(*Note:* This is for aggregates graded mainly between the 19-mm and 9.5-mm size. Sample requirements for other aggregate gradations—38.1 mm to 9.5 mm, 9.5 mm to 4.75 mm, and 4.75 mm to 2.36 mm—are given in the ASTM and CSA standards.)

Procedure:

1. Wash, dry and obtain mass of the sample.
2. Place in the abrasion machine.
3. Add 11 standard steel balls.
4. Rotate the drum for 500 revolutions at 30–33 rpm.
5. Remove the sample. Sieve on a 1.70-mm sieve. Wash the sample retained. Obtain the mass.

Results:

Mass of original sample	_____	g (A)
Mass of final sample	_____	g
Loss	_____	g (B)

Calculations:

$$\% \text{ loss} = B/A \times 100 = \underline{\hspace{3cm}} \%$$

4–7.5 Soundness Test

Purpose: To measure the resistance of aggregates to cycles of freezing and thawing.

Theory: Certain aggregates tend to break up when subjected to cycles of freezing and thawing. Water soaks into pores in the particles; freezes, expanding about 10%; and opens the pores even wider. On thawing, more water can seep in, further widening the crack. After a number of cycles, the aggregate may break apart, or flakes may come off of it. This leads to disintegration of concrete and to weakening of base course layers.

In the soundness test, aggregates are soaked in a solution of $MgSO_4$ or $NaSO_4$. The salt solution soaks into the pores of the aggregate. The sample is removed from the solution, drained, and then dried. During drying, crystals form in the pores, just as ice crystals form in aggregates exposed to weathering. This soaking and drying operation is carried on for a number of cycles. At the end of the test, the amount of material that has broken down is found, and the percentage loss is calculated. (The reliability of this test to simulate freezing and thawing damage is questionable.)

Apparatus: saturated solution of $MgSO_4$
containers for soaking samples
sieves
balance (accurate to 0.01 g)

Procedure:

1. Wash, dry, and obtain mass of test sample [approximately 1000 g if size range is 19–9.5 mm (3/4-3/8 in)].
2. Place in solution for 16–18 hours.

3. Remove, drain, and place in oven for about six hours.
4. Remove when dry. Cool.
5. Repeat steps 2, 3, and 4 for five cycles.
6. Wash the sample thoroughly. Dry.
7. Sieve the sample over an 8-mm (5/16-in) sieve, and measure mass retained.

Results:

Original mass   _____ g (A)
Final mass   _____ g
Loss   _____ g (B)

Calculation:

$$\% \text{ loss} = B/A \times 100 = \text{_____} \%$$

4–7.6 Gradation Analysis (Combined Aggregates)

Purpose: To obtain a grain size distribution curve for a granular material.

Theory: Aggregates which are composed of both fine and coarse sizes must be split on the 4.75-mm sieve, and the two fractions sieved separately. A large sample must be used to obtain the grain size distribution for the coarse sizes. However, due to the large quantity of particles that would be retained on each of the fine sieves, the shaking process would have to be continued for a long time to ensure that all the particles that could pass the fine sieves had done so. Therefore, the fine part of the sample is split down to the amount required for a representative test, about 500–600 g.

Apparatus: sieves
             sieve shaker
             balance (accurate to 0.01 g)

Procedure:

1. Dry and obtain the mass of the test sample.
2. Split on a 4.75-mm sieve.
3. Sieve the coarse sizes, adding the material that passes 4.75 mm to the fine fraction obtained in step 2.

SIEVE ANALYSIS—COMBINED AGGREGATES DATA SHEET

Total sample	Total	Coarse	Fine
Mass dry + pan			
Mass pan			
Mass dry sample	**(a)**		

Passing 4.75 mm (sieving)	+

Total passing 4.75 mm	**(b)**	**(c)** %

$$= \frac{b}{a} \times 100$$

Coarse fraction

Passing sieve	Retained on sieve	Mass retained	% retained		Cum. % passing
	4.75				
4.75	Pan	**(b)***			
	Totals				

*From **(b)** above

Fine fraction
— Washing

	Total	Washed
Mass dry + pan		
Mass pan		
Mass dry sample		

Passing 75 μm

Washed _____

Sieved + _____

Total _____ **(d)**

— Sieving

Passing sieve	Retained on sieve	Mass retained	% retained	Corr. % retained**	Cum. % passing
	75 μm				
75 μm	Pan	**(d)***			
	Totals				

*From **(d)** above

**% Retained x $\dfrac{c}{100}$

4. Split the fine fraction to about 500–600 g. Measure the mass.
5. Wash the sample over a 75-μm sieve. Dry. Obtain the mass.
6. Sieve through fine sieves.

Results: Data can be recorded on the data sheet on page 162.

Calculations: To obtain the corrected percentage retained on each of the fine sieves, multiply the percentage retained on each size by the percentage of the total sample that passes 4.75 mm; for example, if 62% passes 4.75 mm, multiply all values for percentage retained in the fine aggregate sieve analysis by 0.62.

4–8 PROBLEMS

4–1. Define *aggregates.*

4–2. What materials are used in aggregates, and what are the main sources?

4–3. What is meant by coarse aggregate, fines, crushed gravel, crushed rock, and concrete sand?

4–4. What is the nominal maximum size for fine aggregate? What is the actual maximum size usually specified?

4–5. Results of a sieve analysis on an aggregate are:

Pass 50 mm	100 %
Pass 38 mm	93.8%
Pass 25 mm	47.1%
Pass 19 mm	6.1%
Pass 9.5 mm	1.8%

What is a) the nominal maximum size, b) the nominal size range, and c) the term used to describe this size aggregate.

4–6. What size restrictions would you expect to find in the specifications for a 9.5–4.75-mm aggregate? Would this be a coarse or a fine aggregate?

4–7. Why is gradation of aggregates important?

4–8. What is the purpose of a washed test?

4–9. Give two reasons why excessive amounts of fines may be undesirable in aggregates.

4–10. What type of gradation curve is desirable for an aggregate to be used as a highway base course? Why?

4–11. Following are results of a washed sieve analysis:

Original mass   = 608.5 g
Dry mass after washing = 578.2 g

Sieve test:

Sieve	Mass Retained
9.5 mm (3/8 in)	0.0 g
4.75 mm (No. 4)	96.2 g
2.36 mm (No. 8)	117.1 g
1.18 mm (No. 16)	128.8 g
600 μm (No. 30)	105.3 g
300 μm (No. 50)	82.7 g
150 μm (No. 100)	29.3 g
75 μm (No. 200)	14.7 g
Pan	2.7 g

Complete the grain size distribution calculations and draw the grain size distribution curve.

4–12. Results of a sieve test on a coarse aggregate are:

Retained on 25 mm	0 g
19	416 g
12.5	3143 g
9.5	2617 g
4.75	2490 g
2.36	173 g
Pan	87 g

Original mass of the test sample was 8931 g. Calculate the grain size distribution for this aggregate and plot the distribution curve. Is the size of the sample adequate for a sieve analysis according to the ASTM requirements? What is the nominal size of this aggregate?

4—13. In a relative density test on an aggregate sample, the following are recorded:

Dry mass	2117.1 g
SSD mass	2144.3 g
Net volume	786.8 cm^3

Find the bulk relative density.

4—14. Compare aggregate conditions when calculating the apparent relative density and the saturated, surface-dry relative density.

4—15. In a relative density test, the following values are recorded:

SSD mass	2034.2 g
Submerged mass	1276.1 g
Dry mass	2017.1 g

Find the relative density values and absorption.

4—16. An aggregate has an apparent relative density of 2.685. Absorption is 1.2%. Find the bulk relative density.

4—17. What test is used to measure hardness of aggregates? Why is this property important?

4—18. The following results are obtained in an abrasion test on an aggregate:

Original mass	5008.7 g
Final mass	2764.9 g

What are the test results, expressed in the usual terms?

4—19. What may happen if aggregates are exposed to cycles of freezing and thawing? Why? How?

4—20. What test is used to measure resistance to cycles of freezing and thawing? How is it conducted? How are results obtained? Give an example of calculation of the results.

4—21. A soundness test is conducted on a coarse aggregate. The original mass of the sample was 2125 g. After completion of the test, the dry mass of the particles that have not broken down is found to be 1849 g. Find the percentage loss.

4—22. Name two deleterious materials in aggregates and indicate why they are deleterious.

4–23. Name five properties of aggregates and the tests that are used to evaluate each.

4–24. Gradation specifications for a coarse aggregate are:

Sieve	Passing
25 mm (1-in.)	100%
19 mm (3/4-in.)	90–100%
9.5 mm (3/8-in.)	40–65%
4.75 mm (No. 4)	0–15%
2.36 mm (No. 8)	0–5% .

Results of a sieve analysis are:

Sieve	Mass Retained
25 mm (1-in.)	0 g
19 (3/4-in.)	361 g
12.5 (1/2-in.)	1742 g
9.5 (3/8-in.)	1419 g
4.75 (No. 4)	2116 g
2.36 (No. 8)	832 g
1.18 (No. 16)	173 g
Pan	22 g

Does this aggregate meet the specifications? Show the specifications and sample gradation on a grain size distribution graph.

4–25. A highway department includes the following requirements in their specifications for fine aggregate for concrete:

a) Gradation—Pass 9.5 mm   100%
              4.75 mm   95–100%
              2.36 mm   80–100
              1.18 mm   50–85
              600 μm   25–65
              300 μm   10–30
              150 μm   2–10

b) Loss in soundness test, 18% maximum.

c) Amount of clay lumps, 1.0% maximum.

d) Material finer than 75 μm (as measured by the amount lost in washing over the 75-μm sieve), 3.0% maximum.

Results of tests on a possible aggregate are:

i) Sieve analysis
    Sample tested   580.1 g
    Mass after washing   572.2 g
    Results of dry sieving:

Sieve	Mass Retained
9.5 mm	0 g
4.75 mm	21.5 g
2.36 mm	68.1 g
1.18 mm	92.8 g
600 μm	128.7 g
300 μm	106.4 g
150 μm	88.3 g
75 μm	61.4 g
Pan	4.3 g

ii) Soundness test
     Total sample   501.4 g
     Final mass   437.9 g

iii) Test for clay lumps
      Total sample   196.3 g
      Amount of clay lumps   3.5 g

Check this aggregate for acceptance according to the specifications.

4–26. A highway department includes the following requirements in its specifications for 19-4.75-mm coarse aggregate to be used in asphalt pavement.

a) Gradation—Pass 25 mm   100%
                19 mm   90-100%
                9.5 mm   25-60%
                4.75 mm   0-10

b) Material finer than 75 μm, (as measured in a washing test), 1.5%
    maximum.

c) Amount of crushed particles, 60% minimum.

d) Amount of flat or elongated particles, 20% maximum.

Results of laboratory tests on an aggregate being considered for this use
are:

i) Sieve analysis
    Total sample   8226 g

Sieve	Mass Retained
25 mm	0 g
19 mm	622 g
12.5 mm	2955 g
9.5 mm	2619 g
4.75 mm	1844 g
Pan	378 g

ii) Washing test
   Total sample   1981.3 g
   Mass after washing   1942.6 g

iii) Crushed particle test
   Total sample   1262 g
   Amount crushed   677 g

iv) Flat and elongated particles
   Total sample   1719 g
   Amount of flat sand elongated particles   215 g

Check this aggregate for acceptance according to the specifications.

4-27. What factors govern the size of a sample chosen for a gradation test? Why?

4-28. A combined aggregate sample is split on a No. 4 sieve, sieved through the coarse sieves, split down to an acceptable size, washed, and sieved through the fine sieves. Calculate grain size distribution, draw the curve, and check for acceptance as a base course material according to the specification given in section 4-4.1.

Original sample	12387 g
Mass of coarse fraction	7524 g
Mass of fine fraction	4860 g

Coarse sieving:

Retained on 25 mm (1-in.)	0 g
19 mm (3/4-in.)	377 g
12.5 mm (1/2-in.)	1399 g
9.5 mm (3/8-in.)	2643 g
4.75 mm (No. 4)	2956 g
Pan	144 g

Fine fraction split down:

Original sample	533.7 g
Washed sample	512.7 g

Fine sieving:

Retained on 4.75 mm (No. 4)	0.0 g
2.36 mm (No. 8)	105.9 g
1.18 mm (No. 16)	126.6 g
600 μm (No. 30)	77.4 g
300 μm (No. 50)	105.1 g
150 μm (No. 100)	56.8 g
75 μm (No. 200)	32.5 g
Pan	7.2 g

4–29. Following are results of gradation tests on three aggregates:

	Aggregate A	Aggregate B	Aggregate C
Passing 25 mm (1-in.)	100%		
19 mm (3/4-in.)	92%		
9.5 mm (3/8-in.)	41%	100%	100%
4.75 mm (No. 4)	19%	77%	96%
2.36 mm (No. 8)	7%	60%	79%
600 μm (No. 30)	4%	42%	40%
300 μm (No. 50)	2%	36%	16%
75 μm (No. 200)	1%	28%	3%

Combine these to give a gradation falling approximately in the center of the specification given in section 4–4.1.

4–30. The gradation of two aggregates, A and B, are shown below:

	A	B
Passing 25 mm	100 %	—
19 mm	96.4%	—
12.5 mm	48.5%	100 %
9.5 mm	12.7%	91.7%
4.75 mm	2.9%	13.0%
2.36 mm	—	4.2%

These aggregates are to be combined in a 50-50 mixture for use in an asphalt mix. Find the resulting gradation. Check for acceptance according to the specifications quoted in problem 4—26.

5

Pavement Design
and Materials

Pavements are usually surfaced by asphalt mixtures or portland cement concrete, discussed in chapters 6 and 7. However, preparation for surfacing involves construction of the subgrade and a base course. Design of these components depends on the materials to be used and the conditions which the pavement must meet.

5-1 PAVEMENT DESIGN

Pavement types, components, and thickness design procedures are discussed here to illuminate the role of various types of paving material and the soils investigation techniques required for pavement design.

5-1.1 The function of the pavement structure is to distribute imposed wheel loads over a large area of the natural soil. If vehicles were to travel on the natural soil itself, shear failures would occur in the wheel path in most soils and ruts would form. The shear strength of the soil is usually not high enough to support the load. In addition to its load distribution function, the surface course of a highway or airport pavement structure must provide a level, safe travelling surface.

Pavements are classified as "rigid" or "flexible," depending on how they distribute surface loads (fig. 5−1). Rigid pavements are surfaced by portland cement concrete slabs. These act as beams, and distribute the wheel loads fairly

171

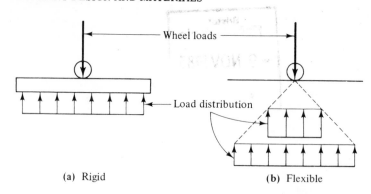

FIG. 5–1. Types of pavement.

uniformly over the area of the slab. Flexible pavements distribute the load over a cone-shaped area under the wheel, reducing the imposed unit stresses as depth increases. The rate of stress reduction varies with the properties of the layers, and is difficult to estimate accurately. However, assuming a 45° cone below the wheel, a tire pressure of 630 kPa (90 psi) at the surface is reduced to 27 kPa (4 psi) at the depth of 400 mm (16 in).

5–1.2   Major components of a pavement structure are:

1. Surface
2. Base   } Pavement
3. Subbase

4. Compacted subgrade }
5. Natural subgrade   } Subgrade

In rigid pavements the surface may be portland cement concrete; in flexible pavements the surface may be asphalt concrete, stabilized or bound granular material, or granular material only.

Typical pavement structures are shown in fig. 5–2.

Bases and subbases are usually granular material or aggregates. The subbase, which is lower in the structures, does not require so high-quality a material as the base.

The compacted subgrade may be (1) the surface layer of the subgrade, compacted in cut areas, or (2) the embankment material in fill zones.

5–1.3   The main function of the pavements is to reduce the high unit stresses imposed by vehicles to stresses on the subgrade that are low enough to be carried without failure due to rutting, excessive settlement, or other types of distress. The magnitude of the stress reduction is mainly a function of the thickness of

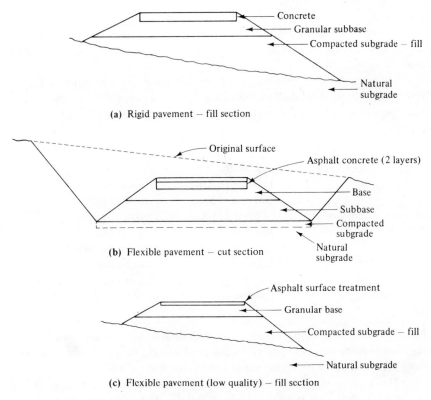

(a) Rigid pavement — fill section

(b) Flexible pavement — cut section

(c) Flexible pavement (low quality) — fill section

FIG. 5—2. Typical pavement structure.

the pavement structure. Therefore, the main variable in the design of the pavement structure is the thickness. Criteria involved in design of the pavement thickness are:

1. The magnitude of imposed loads
2. The strength of the subgrade soil.

Many different methods of measuring the imposed load, the subgrade strength values, and the required pavement structure have been suggested and used. Among pavement design organizations, there is little agreement on the best method. The problem is complex. Wheel loads vary from light passenger cars to heavy transports with dual tandem wheel configurations. Load applications vary from a few thousand to millions per year. Pavements may be designed for various lengths of time. Economics must be considered—whether it is less expensive to construct lower-strength pavements and reinforce them frequently, or to construct relatively maintenance-free highways at high original costs. Soils may vary considerably along a proposed highway route, but the pavement structure cannot be changed continually because of the construction problems that would result.

Many authorities involved with specifications for pavement structures have standardized their designs, based on their experience with structures that have performed adequately in the past. For example, a city design standard might be:

Arterial roads	75-mm (3-in)	asphalt concrete
	150-mm (6-in)	granular base
	300-mm (12-in)	granular subbase
Local roads	40-mm (1½-in)	asphalt concrete
	150-mm (6-in)	granular base
	150-mm (6-in)	granular subbase
Rural roads	150-mm (6-in)	granular base with an asphalt seal coat
	150-mm (6-in)	granular subbase

These are typical examples only. Actual design standards vary considerably, depending on the types of material locally available and the experience the design department has had with maintenance and failure of local roads.

However, many larger pavement design organizations—such as state and provincial highway departments, airport construction agencies, military construction departments, and associations representing engineers or manufacturers of paving materials—have suggested thickness design formulas so that the pavement structure may be more closely related to the type of load to be imposed and the quality of the subgrade.

There are numerous methods of measuring the imposed load—AADT (average annual daily traffic), maximum wheel loads allowed, number of trucks and buses using roads, EWL (equivalent wheel loads), DTN (design traffic number the average daily number of equivalent 18000 lb (80 kN) single axle loads), and others.

Methods used to measure soil quality include those based on soil classification, soil index properties, and soil strength tests of various types.

5–1.4   Methods used to estimate soil strength include:

1. *Group Index.* The group index is an indication of the silt and clay content of a soil. It is calculated as follows:

$$GI = 0.2a + 0.005\,ac + 0.01\,bd \qquad (5-1)$$

where a = that portion of the percentage passing the No. 200 sieve that is greater than 35% and does not exceed 75% (expressed as a positive whole number from 0 to 40);

b = that portion of the percentage passing the No. 200 sieve that is greater than 15% and does not exceed 55% (expressed as a positive whole number from 0 to 40);

c = that portion of the numerical liquid limit that is greater than 40 and that does not exceed 60 (expressed as a positive whole number from 0 to 20); and

d = that portion of the numerical plasticity index that is greater than 10 and that does not exceed 30 (expressed as a positive whole number from 0 to 20).

Example 5–1: Following are the index properties for a soil:

Passing No. 200 sieve—61%

$$w_L = 35$$
$$I_P = 22$$

therefore

$$a = 26 \ (61 - 35)$$
$$b = 40 \ (61 - 15; \text{maximum is } 40)$$
$$c = \ \ 0 \ (35 - 40; \text{minimum is } 0)$$
$$d = 12 \ (22 - 10)$$

and

$$GI = (0.2 \times 26) + (0.005 \times 26 \times 0) + (0.01 \times 40 \times 12)$$
$$= 5.2 + 0 + 4.8 = 10.0$$

2. *California Bearing Ratio (CBR).* Originally developed by the California Division of Highways, this test has been further developed by others and is the most common strength test conducted on soils for evaluation of subgrade quality. Briefly, the test consists of:

   a. Compacting a sample at its optimum moisture content.

   b. Applying a surcharge to the sample to represent the estimated thickness of pavement over the subgrades.

c. Soaking the sample for four days.

d. Forcing a 19.4 cm^2 (3 in^2) plunger into the sample to a depth of 2.5 mm (0.1 in.). The force required to obtain this penetration is expressed as a percentage of the standard load for crushed road base material (13.3 kN or 3000 lb) to give the CBR value.

Example 5–2: In a CBR test, the force required to penetrate the sample 2.5 mm with the piston was 0.85 kN. Therefore the CBR value is 0.85/13.3 = 6.4% = 6.4.

3. *Modulus of subgrade reaction (K).* This is calculated from a field test utilizing a 75-cm (30-in) diameter plate on the subgrade, small plates on this large plate, and a jack to lift a load provided by a loaded transport over the plates. Deflection of the plate is measured for various loads. The modulus of subgrade reaction is the load in pounds per square inch per inch of deflection.

4. *Resistance value (R).* This is a test value resulting from a soil test in a Hveem stabilometer.

5–1.5 Design charts are available for using the above soil quality determination and traffic loading data to estimate required thickness of pavement construction. Figure 5–3 is a thickness design chart based on the group index.

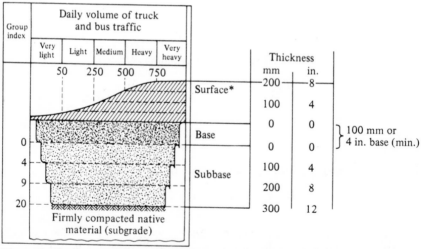

*For surface layers over 75 mm or 3 in. in depth, up to 50% of the asphalt surface may be replaced by twice the thickness of additional granular base.

FIG. 5–3. Thickness design chart–group index (modified).

Example 5–3: A soil has a GI value of 8. Traffic is medium, that is, 250 trucks and buses per day. Select the pavement thickness.

From fig. 5–3, the pavement required consists of:

> 90 mm (3½ in.) of surface
> 100 mm (4 in.) of base
> 200 mm (8 in.) of subbase

5–1.6 The Asphalt Institute recommends "full depth" asphalt pavement structures without any base or subbase layers. Figure 5–4 gives their thickness design recommendations based on CBR or plate bearing values for the soil strength, and the DTN for the traffic load.

Included in fig. 5–4 is the Institute's recommendations for the minimum thickness of asphalt pavement required in cases where bases or subbases are to be utilized. They suggest that additional thickness of asphalt pavement, above the minimum required, can be replaced by base or subbase material using the following equivalencies:

> 2 in or mm of base = 1 in or mm of asphalt
> 2.7 in or mm of subbase = 1 in or mm of asphalt

Example 5–4 Soil along a proposed highway location has a CBR of 3. The DTN is 40. Design a pavement structure using (a) full depth asphalt, (b) asphalt and a base course, and (c) asphalt and a subbase course.

a) From fig. 5–4 (a), the total depth of asphalt pavement required is 10 in or 250 mm.

b) From fig. 5–4 (b), the minimum thickness of asphalt pavement required is 4.6 in. Therefore use 5 in or 125 mm of asphalt. The balance of 5 in can be replaced by 5 × 2.0 = 10 in or 250 mm of base.

c) From fig. 5–4 (b), the minimum thickness of asphalt pavement required is 7.2 in. Therefore use 7.5 in or 190 mm of asphalt. The balance of 2.5 in can be replaced by 2.5 × 2.7 = 7 in or 175 mm of subbase material.

5–1.7 A large-scale test road was constructed and evaluated by AASHTO (formerly AASHO) in the late 1950s. The main steps in the design method that was evolved (for flexible pavements) are, briefly:

1. The soil support value S is found from the group index, CBR, or other test. Conversion to the S value can be made by a straight line on fig. 5–5.

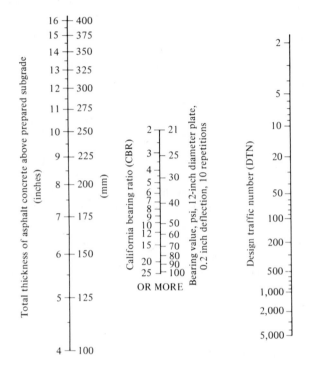

(a) Thickness of asphalt pavement

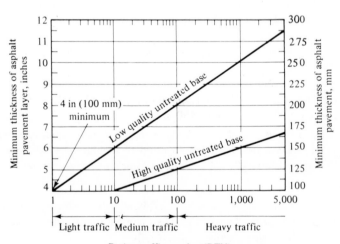

(b) Minimum thickness of asphalt pavement
layers over untreated granular bases

FIG. 5–4. Thickness design chart (courtesy, the Asphalt Institute).

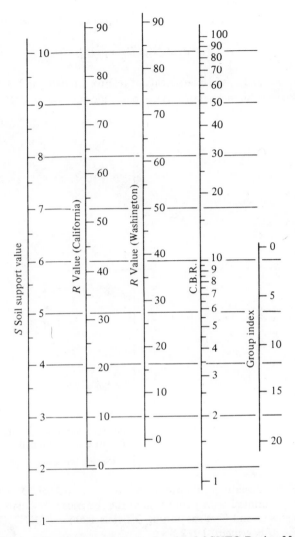

FIG. 5–5. Soil support valve–S–for the AASHTO Design Method. (Reprinted with permission of the American Association of State Highway and Transportation Officials from *AASHTO Guide for Design of Pavement*, 1972.)

2. Obtain (or estimate) the equivalent daily number of 18000-lb (80-kN) load applications.

3. Determine the *structural number*, SN, from fig. 5–6. (Suggested values for regional factors are shown in that figure.)

4. Determine combinations of asphalt surface, base, and subbase from

$$SN = a_1D_1 + a_2D_2 + a_3D_3$$

where a_1, a_2, and a_3 are coefficients of relative strength, and D_1, D_2, and D_3 are thicknesses (in inches) of surface, base, and subbase, respectively.

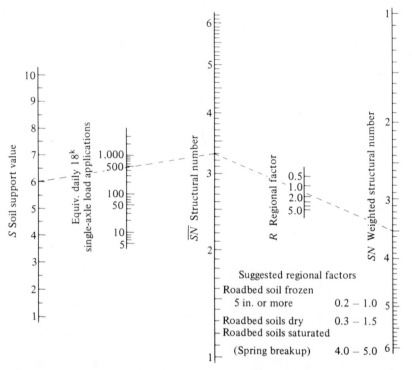

FIG. 5—6. Structural number—SN—for the AASHTO Design Method. (Reprinted with permission of the American Association of State Highway and Transportation Officials from *AASHTO Guide for Design of Pavement*, 1972.)

For more detail, see the AASHTO *Interim Guide for Design of Pavement Structures.*

Suggested values for the strength coefficients are given in table 5—1.

Example 5—5:

A pavement structure to carry 500 80 kN (18000 lb) loads per day is required over a soil with a CBR value of 10. Pavement is to be composed of 100 mm

Table 5-1

SUGGESTED VALUES FOR STRENGTH COEFFICIENTS*

Pavement Component	Coefficient
Surface Course (a_1)	
Roadmix (low stability)	0.20
Plantmix (high stability)	0.44
Sand asphalt	0.40
Base Course (a_2)	
Sandy gravel	0.07
Crushed stone	0.14
Cement-treated (no soil-cement)	
Compressive strength @ 7 days	
650 psi or more (4.48MPa)	0.23
400 to 650 psi (2.76 to 4.48MPa)	0.20
400 psi or less (2.76MPa)	0.15
Bituminous-treated	
Coarse-graded	0.34
Sand asphalt	0.30
Lime-treated	0.15-0.30
Subbase Course (a_3)	
Sandy gravel	0.11
Sandy or sandy-clay	0.05-0.10

* Reprinted by permission of the American Association of State Highway and Transportation Officials from *AASHTO Interim Guide for Design of Pavement*, 1972.

(4 in) of asphalt concrete, 200 mm (8 in) of crushed rock base and the balance a sandy gravel subbase. The subgrade is subjected to frost but is fairly dry. Find the thickness of subbase required.

From fig. 5-5, S = 6.
Assume a regional factor of 1.5.
From fig. 5-6, SN = 3.5.

$$a_1 D_1 = 4 \times 0.44 = 1.76$$
$$a_2 D_2 = 8 \times 0.14 = 1.12$$
$$3.5 - (1.76 + 1.12) = 0.62$$

Therefore

Depth of subbase required = 0.62/0.11 = 5.6 in

or 6 in (rounded to next higher inch) (150 mm)

5–1.8   A simplified approach to thickness design is included in the *Handbook of Highway Engineering*. Traffic load is expressed as the total number of repetitions of an 18000-lb (80-kN) single-axle load that a surface may be subjected to over a 20-year period. Traffic categories for various types of pavement are indicated in table 5–2.

Subgrade categories are "very good," "good," and "poor," and can be estimated from data in table 5–3.

Table 5–2

TRAFFIC CATEGORIES*

I. City Streets and Parking Lots

Designation	Equivalent 18000-lb (80 kN) Single-Axle Loads	Traffic Category
Freeways	30,700,000	11
Local street, industrial	6,570,000	9
Major arterial	5,260,000	9
Local street, business	3,940,000	9
Truck parking lot, entrance	3,500,000	9
Shopping center, truck entrance	1,680,000	9
Collector street	430,000	7
Truck parking lot, parking stalls	35,000	7
Local street, residential	9,500	5
Car parking lot, more than 50 stalls	9,500	5
Car parking lot, 50 stalls or less	950	3
Residential driveways	150	3

II. Rural Roads

Traffic Category	Equivalent 18000-lb (80 kN) Single-Axle Loads	Trucks Per Day In Design Lane
11	above 30,000,000	+4,000
9	+500,000 to 30,000,000	+100 to 4,000
7	+10,000 to 500,000	+5 to 100
5	+1,000 to 10,000	1 to 5
3	less than 1,000	less than 1

* From HANDBOOK OF HIGHWAY ENGINEERING, Robert F. Baker, Editor © 1975 by Litton Educational Publishing, Inc. Reprinted by permission of Van Nostrand Reinhold Company.

Table 5–3
SUBGRADE CATEGORIES*

I. Based on Strength Test

Subgrade Category	California Bearing Ratio (CBR)	Subgrade Modulus K
Very good (VG)	+10	+200
Good (G)	+ 6 to 10	+150 to 200
Poor (P)	3 to 6	100 to 150

II. Based on Soil Classification[a]

Subgrade Category	Material	Unified System	AASHTO System
Very good	Gravels and sandy gravels	GW, GP, GM, GC SW, SP, SM, SC	A-1, A-2-4, A-2-5, A-2-6, A-2-7, A-3
Good or poor	Silts and clays	ML, CL, OL, MH CH, OH	A-4, A-5, A-6, A-7-5, A-7-6

Silts and clays rated poor only under the following conditions:
(a) When they occur in low-lying areas where the natural drainage is very poor and will not be improved.
(b) Where the condition of water table and climate are such that severe frost heave can be expected.
(c) Where high percentages of mica-like fragments or diatomaceous particles produce a highly elastic condition. This would occur mainly in A-5 (ML and MH) soils.
(d) Where it is desired to "bury" highly expansive soils, usually A-7-6 (CH), deeper in the section to limit the effects of seasonal variations in moisture.

* From HANDBOOK OF HIGHWAY ENGINEERING, Robert F. Baker, Editor © 1975 by Litton Educational Publishing, Inc. Reprinted by permission of Van Nostrand Reinhold Company.

Thickness design for full-depth asphalt pavements, asphalt surfaces with bases (and subbases), and concrete pavements are indicated in table 5–4.

5–1.9 A thorough analysis of thickness design methods is beyond the scope of this book. The above methods have been introduced to illustrate the importance of the properties of the subgrade soil and pavement materials in thickness design and to introduce some simple design methods.

In addition to the *AASHTO Interim Guide* referred to in section 5–1.7, comprehensive thickness design methods are covered in "Thickness Design for

Table 5–4
THICKNESS DESIGN*

Thickness, Inches (mm) for Subgrade Class

Traffic Category	Component	Very Good			Good			Poor		
		Full Depth Asphalt	Asphalt Surface	Concrete	Full Depth Asphalt	Asphalt Surface	Concrete	Full Depth Asphalt	Asphalt Surface	Concrete
11	Surface		2.75 (70)			4.0 (100)			6.0 (150)	
	Base		12.0 (300)			14.0 (350)			18.0 (450)	
	Total	8 (200)	14.75 (370)	6 (150)	11 (295)	18.0 (450)	8 (200)	16 (400)	24.0 (600)	10 (250)
9	Surface		2.5 (65)			3.5 (90)			5.0 (125)	
	Base		11.0 (280)			13.0 (320)			16.0 (400)	
	Total	7 (175)	13.5 (345)	6 (150)	9 (225)	16.5 (410)	7 (175)	13 (325)	21.0 (525)	9 (225)
7	Surface		2.0 (50)			3.0 (75)			4.0 (100)	
	Base		10.0 (250)			12.0 (300)			14.0 (350)	
	Total	6 (150)	12.0 (300)	6 (150)	8 (200)	15.0 (375)	6 (150)	11 (275)	18.0 (450)	8 (200)

5	Surface	5 (125)	5 (125)	6 (150)	6 (150)	8 (200)
	Base		1.75 (45) 8.0 (200)	2.0 (50) 10.0 (250)		2.75 (70) 12.00 (300)
	Total	5 (125)	9.75 (245)	12.0 (300)	6 (150)	14.75 (370) 6 (150)
3	Surface	4 (100)	5 (125)	5 (125)	7 (175)	
	Base		1.5 (40) 6.0 (150)	1.75 (45) 8.0 (200)	2.5 (60) 11.0 (280)	
	Total	4 (100)	7.5 (190)	9.75 (245)	5 (125)	13.5 (340) 6 (150)

* From HANDBOOK OF HIGHWAY ENGINEERING, Robert F. Baker, Editor © 1975 by Litton Educational Publishing, Inc. Reprinted by permission of Van Nostrand Reinhold Company. (Mm values for pavement thickness were added by the author.)

Notes:

- Subbases are usually used under concrete pavements for Traffic Categories 7, 9, and 11. Subbases are also frequently used to replace part of the base requirements, particularly when base thicknesses exceed 8 inches or 200 mm. Authorities do not agree on the desirability of subbases under full-depth asphaltic concrete; some engineers believe subbases are needed for some soil conditions, particularly for Traffic Categories 7, 9, and 11; others believe they are detrimental. The use of subbases and soil and base stabilization is based on experience, and whenever their use is contemplated, advice from area highway engineers should be sought.

- Designs based on this table will produce good results for Traffic Categories 3 and 5. For Traffic Categories 7, 9, and 11, the table will provide a reasonable estimate of cost and general requirements. If experienced designers are used to review (or preferably design) the thicker pavements, the probability of more economic construction and maintenance is increased.

- Local experience is a major requirement for pavement design because of the combined effect of soil, materials, and the environment. Furthermore, there are differences in opinion on minimum thickness of asphaltic concrete, the need to vary surface thickness with subgrade type, etc. These differences are frequently nominal, and are unlikely to lead to grossly different costs and thicknesses than are provided in the table.

Concrete Pavements" and "Design of Concrete Pavements for City Streets" published by the Portland Cement Association, and in "Thickness Design–Asphalt Pavements for Highways and Streets," published as Manual MS–1 by the Asphalt Institute.

5–2 SUBGRADE CONSTRUCTION

Subgrade construction, from the materials point of view, consists mainly of:

1. Compaction of the top layer of subgrade in cuts, and the whole depth of the added material in fills; and
2. Identification and treatment of unsuitable material.

Soil compaction was discussed in detail in chapter 3. AASHTO-recommended minimum requirements for compaction of embankments and subgrades are given in table 5–5. In general, they require 95% compaction of embankment material and 100% compaction of the surface layer of subgrade, usually 150 mm (6 in).

Unsuitable material includes organic material and frost-susceptible soils (in areas subject to freezing temperatures).

5–2.1 Topsoil is the surface layer of most existing soils. Due to its high organic content, it is quite compressible. It is usually removed in fill areas, when it is located within a few feet of the final subgrade surface.

Table 5–5

RECOMMENDED MINIMUM COMPACTION REQUIREMENTS*

Type of Soil (AASHTO Classification)	Compaction Requirements (% of density obtained in the Standard Compaction Test)		
	Embankments under 50 ft (15 m)	Embankments over 50 ft (15 m)	Subgrade (immediately under pavement structure)
A-1, A-2-4, A-2-5, A-3	95	95	100
A-2-6, A-2-7, A-4, A-5, A-6, A-7	95	**	95†

* Adapted from AASHTO Designation M57 by permission of the American Association of State Highway and Transportation Officials from *AASHTO Materials,* 12th edition, 1978.

** Use of these materials requires special attention to design and construction.

† At 95% of optimum moisture content.

Other organic deposits are found in swamps and muskeg areas. In these cases the soil is composed mainly of organic material and is very highly compressible. Treatment may include:

1. Floating the pavement.
2. Excavation and replacement of the organic material.
3. Displacement of the organic deposit using a surcharge. (Displacement may be aided by water jetting or explosive charges to force the organic material to the sides of the pavement, allowing the surcharged pavement structure to fall into the void remaining.)
4. Use of geotextiles to separate base material from the subgrade and to reinforce subgrade strength (see section 5–5).

Construction techniques vary widely among pavement construction authorities.

Identification of the soil types is fairly simple, and has already been discussed.

5–2.2 Frost damage to pavement is the most serious cause of pavement structure failure in climates subject to freezing conditions. The common occurrence known as "spring breakup" in low-class roads is caused mainly by frost action in susceptible soils.

Capillary rise in soils is a major factor in frost damage, as mentioned in section 1–5.7. Water rises in capillary tubes above the water table by a distance that varies inversely with the average size of the pores in the soil structure.

The mechanics of the growth of ice lenses have been discussed, but are briefly outlined again here.

As a freezing front descends in the soil during the fall and winter, the water in the larger pores freezes. However, the capillary water in adjacent, smaller pores does not freeze, due to the depression of the freezing temperature in these very small volumes of water. This super-cooled water moves to the previously formed ice crystals and freezes on the crystal. If the capillary water is replaced while the freezing front remains stationary, an ice lens builds up in the soil. Under conditions favorable to lens development, the ice lens—consisting mainly of excess water brought up from the groundwater table—may grow to 5-10 cm (2-4 in) or more in thickness. As the freezing front penetrates farther into the soil, more ice lenses are formed and cause a heave in the road surface. Soil heaves of more than 30 cm (1 ft) have occurred. Vertical cuts in these heave areas reveal a series of ice lenses made up of relatively clear ice, with the total thickness equalling the amount of heave. Figure 1–22 illustrates conditions leading to growth of ice lenses and frost heaving.

During thawing of a heaved soil, pavement breakup may occur, as illustrated in fig. 5—7. "Frost boils" or wet potholes are formed on the surface as the roadbed thaws due to the excess water and loose condition of the roadbed.

For frost damage to occur, these conditions are required:

1. Freezing temperatures.
2. A source of free water.
3. A frost-susceptible soil.

The frequency and duration of freezing temperatures vary according to the local climate. The amount of pavement damage in susceptible locations varies with the depth of frost penetration. Damage may also increase if the soil is subjected to cycles of freezing and thawing during the winter.

The source of free water is usually the groundwater, although seepage from higher areas may also feed ice lenses.

Susceptible soils are those that (1) have a fairly high capillary rise, and (2) are permeable enough to allow the transport of a sufficient quantity of water to the ice lenses during the time that the freezing front is stationary. Ice lenses can ordinarily occur only in the zone of the capillary fringe, as it is the capillary water that feeds the ice lens and allows it to grow. Also required is a rate of water movement in the soil that allows a quantity of water to be transported to the ice lens that is sufficient to build a damaging thickness before the freezing front descends farther. Therefore frost-susceptible soils include silts, silty clays, very fine sands, and sands and gravels containing silt or clay fines. Clean sands and gravels have very low capillary fringes [$hc = 30$ cm (1 ft) or less]; therefore

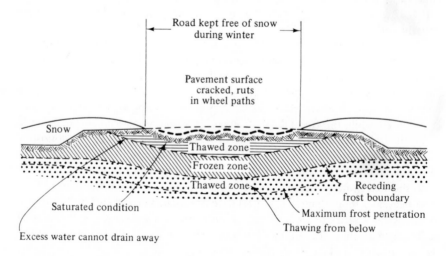

FIG. 5—7.   Spring break-up.

few, if any, ice lenses can form. Fairly impervious soils, such as heavy clays, are not particularly frost-susceptible unless they contain seams of more permeable materials, such as in varved clay. These conditions are illustrated in fig. 5–8.

Identification of frost-susceptible soils is extremely difficult. Involved are the amount of capillary rise and the permeability of the soil, which vary mainly with the pore size, and therefore vary with the grain size, grain type, grain size distribution, and density of the soil.

The most commonly quoted table for identification of frost-susceptible soils is that of the U.S. Army Corps of Engineers (table 5–6). It is based mainly

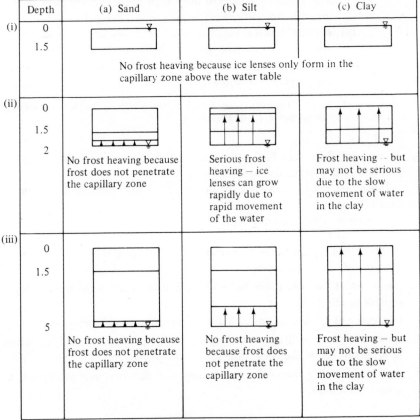

Note: Based on following typical values—
        Depth of frost penetration: 1.5 m
        Height of Capillary rise: sand, 0.2 m; silt, 1.5 m; clay, 5.0 m
        Water table: (i) 0 (at surface); (ii) 2 m; (iii) 5 m

FIG. 5–8. Conditions leading to frost heaving.

Table 5–6

FROST-SUSCEPTIBLE SOILS*

Group	Description
F1	Gravelly soils containing between 3 and 20 percent finer than 0.02 mm by weight.
F2	Sand containing between 3 and 15 percent finer than 0.02 mm by weight.
F3	(*a*) Gravelly soils containing more than 20 percent finer than 0.02 mm by weight, and sands, except fine silty sands, containing more than 15 percent finer than 0.02 mm by weight. (*b*) Clays with plasticity indices of more than 12. (*c*) Varved clays existing with uniform conditions.
F4	(*a*) All silts including sandy silts. (*b*) Fine silty sands containing more than 15 percent finer than 0.02 mm by weight. (*c*) Lean clays with plasticity indices of less than 12. (*d*) Varved clays with nonuniform subgrade.

* Originally published in "Engineering and Design, Pavement Design for Frost Conditions," U.S. Army Corps of Engineers, EM-1110-345-306.

on the amount finer than 0.02 mm in the soil. According to its classification, groups increase in possible damage from F1 to F4.

Table 5–7 roughly indicates the possible frost susceptibility of soils based on classification.

Techniques used to control frost damage include:

1. Removal and replacement of frost-susceptible material within all or part of the zone of frost penetration.
2. Insulation of the susceptible soil with rigid foam sheets to reduce the depth of frost penetration.
3. Deepening of ditches or provision of subdrains to lower the water table.
4. Construction of capillary cutoff layers, made up of coarse sands or waterproof sheets below the susceptible soil.
5. Construction of a thicker, stronger pavement structure over susceptible or questionable soils.

5–3   GRANULAR BASE COURSES

Base courses in pavement structures are composed of either solely granular materials (aggregates), or soil or granular materials stabilized by an additive.

Table 5–7

POTENTIAL FOR FROST ACTION OF ENGINEERING
SOIL CLASSIFICATION GROUPS*

| | Soil Classification Groups | |
Potential Frost Action	Unified	AASHTO
None to very slight	GW, GP, SW, SP	A-1, A-3
Slight to medium	GM, GC	A-2
Slight to high	SM, SC	A-2
Medium	CH, OH	A-7
Medium to high	CL, OL	A-6
Medium to very high	ML, MH	A-4, A-5

* From HIGHWAY MATERIALS by Krebs and Walker. Copyright © 1971 by McGraw-Hill, Inc. Used with permission of McGraw-Hill Book Company.

Granular base courses are mainly aggregates from sand or gravel deposits or from quarries. Slag and other materials are also used in some areas.

Properties required in these materials vary with the type of pavement and the depth of the material in the pavement structure.

5–3.1 Base courses in flexible pavements must help to distribute the load. This ability to distribute the load is primarily a function of the depth of the base course. The quality of material in the base course also affects rate of distribution to a certain extent, but depth has been the main factor considered in design. While distributing the load, the base course itself must not be a cause of failure. Thus it must be strong enough to carry the load without shear failure and resultant rutting. To ensure that its strength is maintained, the base course must allow water drainage to the sides of the pavement structure. If the base becomes saturated, high stresses may be created in the water occupying the pore spaces, resulting in less frictional strength between particles. Nor can the base be susceptible to frost action, which would seriously deplete its strength during the spring thaw. As these properties are dependent to a certain extent on the grain size distribution in the aggregate and, in particular, on the amount of material passing the 75-μm (No. 200) sieve, the aggregates must be durable, that is, resistant to degradation or break-down to smaller sizes from wear and weathering.

The base course must also prevent infiltration of subgrade material.

Subbases for flexible pavements also must have relatively free drainage and must be free of frost action. Strength is not so important, however, since the course is lower in the pavement structure and therefore withstands much smaller loads. Durability of aggregate particles is also important to prevent degradation.

One granular course is usually placed under rigid pavement surfaces. This is usually called the subbase. The subbase's main function is to prevent *pavement pumping,* which has been the most serious cause of failure of rigid pavement bases. The mechanics of pumping are:

1. Creation of a shallow void under the slab due either to slight slab deflection with each load or to curling of the slab due to temperature variations between the top and bottom of the slab.
2. Water filling the void space due to infiltration of rain or other causes.
3. Mixing of water and fines in the underlying material.
4. Ejection of this mixture at joints and edges of the pavement as each wheel load causes the slab to deflect.

Large voids have been created under slabs by pumping. The main function of the subbase is to prevent the pumping that would usually occur if the slab were constructed directly on the subgrade. To prevent pumping, the subbase must be free-draining so that water is readily removed. At the same time it must be free from frost damage, durable enough to prevent degradation, and resistant to infiltration of subgrade soils.

5–3.2 The main quality specification for a granular base course is the grain-size distribution requirement. As noted in chapter 4, the shape of the curve is a good indication of strength. Flexible pavement bases are dense-graded, and therefore fall within limits that result in a curve of roughly maximum density. Strength is not so important for subbases; therefore grading requirements are usually more open. The percentage of fines is usually restricted to a low maximum to ensure that the course is free-draining and free from frost damage. Excessive fines could fill the voids in the mix and greatly reduce permeability; this would also increase the amount of capillary rise, and therefore make the pavement most frost-susceptible.

Other requirements by various authorities include:

1. *Maximum values for plasticity*—to control the amount of clay fines, since these are considered more dangerous as far as frost damage is concerned.
2. *Maximum loss in abrasion test*—to ensure that hard aggregates are used that will not degrade to smaller sizes (thus increasing the fines content) during construction or during the life of the pavement.
3. *Maximum loss in soundness test*—to prevent degradation due to cycles of freezing and thawing.
4. *Petrographic requirements*—to govern overall aggregate quality.

Subbases for both flexible and rigid pavements need not meet the high-strength requirements governing flexible pavement bases, and therefore do not have to be dense-graded. However, open-graded courses may be contaminated by the intrusion of fine-grained subgrade soil, thus reducing their drainability, strength, and resistance to frost damage. To prevent this infiltration when coarse-graded subbase materials are used, the following requirements are often specified:

$$\frac{D_{15} \text{ subbase}}{D_{15} \text{ subgrade}} > 4$$

$$\frac{D_{15} \text{ subbase}}{D_{85} \text{ subgrade}} < 4$$

Example 5–6: A subgrade soil has a gradation which shows that the soil has D_{15} = 0.003 mm and D_{85} = 0.074 mm. The proposed subbase material has D_{15} = 0.14 mm. Does this material meet the criteria to prevent intrusion of subgrade material into the subbase?

$$D_{15} \text{ subbase}/D_{15} \text{ subgrade} = 0.14/0.003 = 47$$
$$D_{15} \text{ subbase}/D_{85} \text{ subgrade} = 0.14/0.074 = 1.9$$

The subbase material meets the criteria.

If the subbase cannot meet these requirements, a thin filter layer that does meet them should be used directly on the subgrade.

For construction purposes, specifications may require a minimum percentage of fines. This is needed to:

1. Ensure that the finished granular surface is stable enough to carry the paving equipment without disruption of the surface.
2. Ensure that the material can be compacted adequately. Some fines are required so that when the subbase is compacted, the granular material will bind and the particles will interlock. Otherwise they would shear or move when loaded.

5–3.3 Typical specifications for granular bases and subbases are given in tables 5–8 and 5–9.

Table 5–8

SPECIFICATIONS FOR BASE AND SUBBASE MATERIALS*

Sieve Designation		Grading Requirements — Mass Percent Passing					
Standard mm	Alternate	Grading A	Grading B	Grading C	Grading D	Grading E	Grading F
50	2 in.	100	100	⋯	⋯	⋯	⋯
25.0	1 in.	⋯	75–95	100	100	100	100
9.5	3/8 in.	30–65	40–75	50–85	60–100	⋯	⋯
4.75	No. 4	25–55	30–60	35–65	50–85	55–100	70–100
2.00	No. 10	15–40	20–45	25–50	40–70	40–100	55–100
0.425	No. 40	8–20	15–30	15–30	25–45	20–50	30–70
0.075	No. 200**	2–8	5–20	5–15	5–20	6–20	8–25

Additional Requirements:

1. Coarse aggregate—not over 50% loss in Los Angeles Abrasion Test.
2. Amount passing No. 200 shall be not more than 2/3 of the amount passing No. 40.
3. Fraction passing No. 40   maximum w_L = 25
   maximum I_P = 6

* Adapted from AASHTO Designation M147 by permission of the American Association of State Highway and Transportation Officials from *AASHTO Materials* 12th edition, 1978.

** The percentage passing No. 200 may be lowered if that is required to prevent frost damage.

Table 5–9

SPECIFICATIONS FOR AGGREGATE MATERIAL FOR BASES OR SUBBASES*

Grading Requirements

Sieve Size	Design Range Percentages Passing	
	Bases	Subbases
2 in. (50 mm)	100	100
1½ in. (37.5 mm)	95 to 100	90 to 100
3/4 in. (19.0 mm)	70 to 92	. . .
3/8 in. (9.5 mm)	50 to 70	. . .
No. 4 (4.75 mm)	35 to 55	30 to 60
No. 30 (600 μm)	12 to 25	. . .
No. 200 (75 μm)	0 to 8**	0 to 12 **

Additional Requirements:

1. At least 75% of particles retained on 3/8-in. sieve shall have at least two fractured faces.
2. Amount passing No. 200 shall be not more than 60% of the amount passing No. 30.
3. Fraction passing No. 40   maximum w_L = 25
   maximum I_P = 4 (6 if below frost)

 * Adapted from ASTM Standard D2940 and reprinted by permission of the American Society for Testing and Materials, 1916 Race Street, Philadelphia, PA 19103, Copyright.
** If frost damage may occur, the maximum allowable passing No. 200 may be lowered, and a maximum limit of 3% smaller than 20 μm is recommended.

5–4 STABILIZED BASE COURSES

Soil and aggregate stabilization allows the use of low-quality aggregates or *in situ* materials as bases (or improved subgrades) in pavement structures. Portland cement, asphalt cement, and lime are the most commonly used stabilizing agents.

5–4.1 Lime is often used with highly plastic, fine-grained soils. Amounts vary from 2% to 8% by weight of soil. When mixed and compacted, the plasticity and swelling potential of the soil are reduced, and workability is increased. Later the soil gains in strength from the cementing properties of the lime.

Asphalt stabilization is used with granular soils, silts, and silty clays. Asphalt strengthens and waterproofs the soil; with granular materials, it can be used to make a high-quality base material. This is also known as *deep strength construction.*

Portland cement often provides an acceptable base with sands, silts, and silty clays. In combination with contaminated road base material, it may be used in reconstruction of streets. It is also used to improve the quality of granular base aggregates. When used with subgrade or old pavement materials, it is commonly called *soil cement*.

5–4.2   Soil cement has provided acceptable bases for many pavements. Since it would deteriorate rapidly if exposed to traffic, it must be surfaced with a wearing course, usually asphalt concrete. A thickness of 150–200 mm (6-8 in) is often used as an acceptable base.

Design of a soil cement course primarily involves the selection of an appropriate cement content. A number of briquettes are made with various cement contents and with optimum moisture contents (as determined by the standard compaction test). These briquettes are then subjected to durability tests—either cycles of freezing and thawing, or cycles of wetting and drying. The cement content specification is designed to ensure that the total abrasion loss after a number of cycles of testing is kept to a maximum, given in table 5–10. Approximate cement contents are indicated in table 5–11.

Example 5–7:   Briquettes (A-2-7 soils) were tested for durability, with the following results.

Test	Sample	Percentage Cement	Percentage Loss
Freeze-Thaw	1	5	13.2
	2	7	11.3
	3	9	8.6
Wet-Dry	1	5	11.9
	2	7	8.5
	3	9	6.7

Find the design cement content.

Plotting the results of loss vs. cement content, these durability curves are obtained:

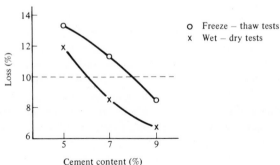

Cement content (%)

Table 5–10

MAXIMUM ALLOWABLE LOSS IN DURABILITY TESTS FOR SOIL CEMENT AS RECOMMENDED BY THE PORTLAND CEMENT ASSOCIATION*

Soil Type	AASHTO Class	Maximum Loss
Sands, gravels	A-1, A-3, A-2-4, A-2-5	14%
Silty soils	A-2-6, A-2-7, A-4, A-5	10%
Clayey soils	A-6, A-7	7%

* Courtesy of the Canadian Portland Cement Association.

A cement content of 8% is required to meet both durability requirements. (Maximum loss is 10%.)

5–4.3 Components of a soil cement mixture and calculations of quantities are as follows:

Air	$M_A = 0$
Water	M_W
Cement	M_C
Solids	M_D

V M

$$\text{Water content } (w) = \frac{M_W}{M_D + M_C} \qquad (5-2)$$

$$\text{Cement content } (cc) = \frac{M_C}{M_D} \qquad (5-3)$$

$$\text{Dry density } (\rho_D) = \frac{M_C + M_D}{V} \qquad (5-4)$$

$$\text{Density } (\rho) = \frac{M}{V} \qquad (1-3)$$

Example 5–8: A soil-cement mixture contains the following quantities per m^3: 1750 kg of dry soil, 140 kg of cement, and 210 kg of water. Find the density, dry density, water content, and cement content.

$$\text{Density} \qquad \frac{2100}{1} = 2100 \text{ kg/m}^3$$

$$\text{Dry density} \qquad \frac{1890}{1} = 1890 \text{ kg/m}^3$$

$$\text{Cement content} \qquad \frac{140}{1750} = 8.0\%$$

$$\text{Water content} \qquad \frac{210}{1890} = 11.1\%$$

Table 5–11

APPROXIMATE CEMENT CONTENTS FOR SOIL CEMENT MIXTURES*

AASHTO Soil Group	Usual Range in Cement Requirement, Percent by Mass	Estimated Cement Content and that Used in Moisture-Density Test Percent by Mass	Cement Contents for Wet–Dry and Freeze–Thaw Tests, Percent by Mass
A-1-a	3– 5	5	3– 5– 7
A-1-b	5– 8	6	4– 6– 8
A-2	5– 9	7	5– 7– 9
A-3	7–11	9	7– 9–11
A-4	7–12	10	8–10–12
A-5	8–13	10	8–10–12
A-6	9–15	12	10–12–14
A-7	10–16	13	11–13–15

* Courtesy of the Canadian Portland Cement Association.

Example 5–9: Following are the design proportions for a soil-cement mixture:

Dry density	2240 kg/m^3
Water content	9.0%
Cement content	7.5%

Find the mass of the dry soil, water, and cement per m^3.

$$M_C + M_D = 2240 \text{ kg}$$

$$cc = 7.5\% = \frac{M_C}{M_D}$$

Therefore

$$M_D = \frac{M_D + M_C}{1 + cc} = \frac{2240}{1.075} = 2084 \text{ kg}$$

$$M_C = 2240 - 2084 \qquad = \quad 156 \text{ kg}$$

$$M_W = 9.0\% \times 2240 \qquad = \quad 202 \text{ kg}$$

$$\text{Total} \qquad\qquad 2442 \text{ kg}$$

Field mix calculations would have to take into account the natural moisture content of the soil.

Example 5–10: A mix design for soil cement (see example 5–9) requires 2084 kg of dry soil, 156 kg of cement, and 202 kg of water per m³ of mix. The soil being used contains 4.0% water. Adjust these requirements to give the mass of the natural soil, cement, and added water required.

$$2084 \text{ kg soil contains } 4\% \text{ water} = 83 \text{ kg water}$$

Field mix proportions are:

$$2084 + 83 = 2167 \text{ kg natural soil}$$
$$202 - 83 = \ \ 119 \text{ kg water}$$
$$\underline{\qquad 156} \text{ kg cement}$$

Total   2442 kg

5–4.4 When cement is used to improve the quality of granular soils, the product is called *cement-modified granular material.* This is used for granular soils of which less than 25% passes the 75-μm (No. 200) sieve.

One method used to assign a minimum cement content is to:

1. Conduct compaction tests with the approximate required cement content (6%–10%, usually).
2. Make a series of compacted samples at a range of cement contents and at the optimum water content determined in (1).
3. Cure the samples for seven days in a humid room.
4. Find the compressive strength (the required compressive strength is that which is specified or that which meets the minimum strength as shown in fig. 5–9).

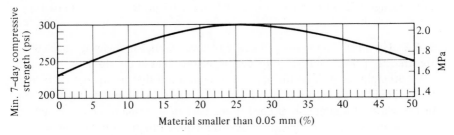

FIG. 5–9. Minimum strength requirements for cement-modified granular materials (no material coarser than No. 4 sieve).

As noted in the section on thickness design, three types of cement-treated bases are recognized for relative strength coefficients—those with compressive strengths over 650 psi (4500 kPa), 400–650 psi (2800–4500 kPa), and less than 400 psi (2800 kPa) in the AASHTO design method.

5–4.5 Construction of soil cement and cement-improved bases involves:

1. Pulverizing soil, if required.
2. Mixing with cement and water.
3. Laying and compacting.
4. Curing, to allow hydration to continue.
5. Surfacing, to protect from surface abrasion.

5–5 GEOTEXTILES

Geotextiles, also called *engineering fabrics* or *filter fabrics,* are permeable synthetic textile products, used in highway, soils, and marine engineering projects.

5–5.1 The fibers used in geotextiles—polypropylene, polyester, polyamide (nylon), and polyethylene—do not rot or decompose in ordinary use. Ultraviolet light results in deterioration but many fabrics are treated with special coating to protect them from sunlight.

Two main types of fabrics are manufactured, depending on how the fibers (yarns) are combined. Woven fabrics are produced, as are most textiles, by the weaving process in which longitudinal fibers are interlaced with crosswise fibers. Nonwoven fabrics are produced by the fibers combining in a random orientation as they are extruded from the production machinery. They are bonded together afterward by heat or other methods so that the orientation of the fibers to each other is permanent.

These fabrics are commonly available in rolls 3.5 m (12 ft) in width and various lengths. Thicknesses vary from 0.3 mm (0.01 in) to over 10 mm (0.4 in).

5–5.2 The most important properties of these fabrics in highway engineering are strength, permeability, and the size of the openings in the cloth.

Strength, as well as the amount of elongation that occurs before failure, is measured by the *grab test* and the *ball burst test.* In the grab test a small part of the width of a sample is gripped in the jaws of a tensile machine and pulled until failure. The ball test measures the load required to puncture the fabric with a ball. Woven fabrics usually have higher strength but with a much lower elongation at failure.

Water must be able to flow through these fabrics easily without allowing any pressure build-up in the water. The fabric must be more permeable than the soil containing the water. A permeability test, similar to the constant head permeability test, is used to measure this property. Results must be used cautiously as the pores or openings in the cloth may become less permeable due to clogging by soil fines.

To be effective in most installations the fabric, while allowing water flow, must prevent the movement of soil particles with the water. The opening size must be small enough to prevent entry of soil particles into the frabic. The *Equivalent Opening Size* (EOS) is approximately the largest opening or pore size in the fabric. The EOS is measured by using the fabric as a sieve and determining the size of glass beads that will go through it under certain conditions.

5–5.3 The main functions of geotextiles in highway projects are separation, reinforcement, and filtration, as illustrated in fig. 5–10.

As a separation device, the fabrics separate weak subgrade soils or organic deposits from aggregate bases. Without separation the subgrade would intrude into the aggregate, resulting in loss of the aggregate as a load-distributing layer. The large aggregate particles end up as isolated pieces of aggregate in the soft soil mass.

The fabric is usually laid directly over the soft or marshy soil. The vegetation helps to spread the load out also.

To allow water flow but not movement of soil particles the EOS of the fabric should be less than the D_{85} of the subgrade soil and permeability of the fabric should be 2 to 10 times that of the subgrade.

In the reinforcement function, the geotextile provides tensile strength to the soil-fabric system.

When used at the boundary between base aggregates and weak subgrades the fabric spreads the wheel-load out and prevents the formation of deep ruts. Tensile stress in the fabric carries some of the load.

In retaining wall construction, fabric can be laid horizontally between layers of soil providing tensile strength to help prevent failure.

Under embankments these fabrics may help to prevent failure along a circular-arc failure plane through the embankment and subgrade. The fabric must tear to allow failure in this case.

High strength at low elongation, suggesting the use of woven fabrics, is usually best for reinforcement.

Geotextiles used as filters protect drains, ditches, and slopes by preventing soil fines from moving into drains or eroding from ditches, along shorelines, and from under water control structures.

Again the requirement that the EOS be less than the D_{85} of the soil is of prime importance. Nonwoven fabrics usually are best for filtration application. Permeability should be at least twice that of the soil being protected.

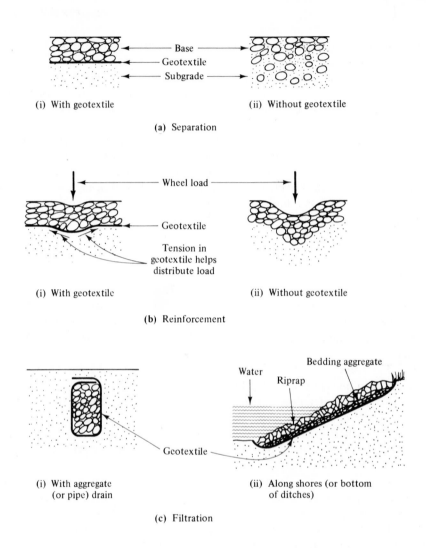

(i) With geotextile   (ii) Without geotextile

(a) Separation

(i) With geotextile   (ii) Without geotextile

(b) Reinforcement

(i) With aggregate   (ii) Along shores (or bottom
(or pipe) drain   of ditches)

(c) Filtration

FIG. 5–10. Uses of geotextiles in highway engineering.

5–5.4 Geotextiles are currently used mainly with sands and organic soils. Gravels normally do not require strengthening or protection against erosion. Fabrics are not available with small enough EOS values to resist infiltration and clogging by clay grains and therefore are not effective in use with these soils when water must be allowed to drain.

5–6 TEST PROCEDURES

Two test procedures are included here. For standard methods for these and other tests on materials for pavement bases, refer to the ASTM and AASHTO standards listed in the Appendix.

5–6.1 California Bearing Ratio
5–6.2 Compressive Strength of Cement-Modified Granular Materials

5–6.1 California Bearing Ratio

Purpose: To measure strength and swelling potential of a soil.

Theory: The California Bearing Ratio test is one of the most commonly used methods to evaluate the strength of subgrade soil for pavement thickness design. A soil is compacted in a mold with the standard compactive effort at its optimum water content (so that it is at about 100% of its maximum density, as determined by the standard compaction test). This test simulates the prospective actual condition at the surface of the subgrade. A surcharge is placed on the surface to represent the mass of pavement materials above the subgrade. The sample is soaked to simulate its weakest condition in the field. Expansion of the sample is measured during soaking to check for potential swelling. After soaking, the strength is measured by recording the force required to shove a penetration piston into the soil.

Apparatus: compression machine with penetration piston that is 49.5 mm
      (1.95 in.) in diameter, with an area of 19.35 cm^2 (3.0 in^2)
    mold that is 152.4 mm (6.0 in.) in diameter and 177.8 mm (7.0 in.)
      high, with collar and base
    spacer that is 61.4 mm (2.416 in.) high to fit mold
    standard compaction hammer
    surchage masses, each weighing 2.27 kg (5 lb)
    swell-measuring apparatus

Procedure:

1. Place the spacer in the mold, with base plate and collar attached.
2. Fill the mold with three layers of soil at its optimum moisture content, compacting each layer with 56 blows of the standard compaction hammer. Remove the collar. Level the surface with a straightedge.
3. Remove the perforated base plate, invert the mold, and replace it on the base plate.

4. Remove the spacer, insert the expansion measuring apparatus, place the surcharge rings on the surface, and take an initial swell-gauge reading.

   Note: A minimum of 4.54 kg (10 lb) of surcharge is used, which represents about 125 mm (5 in) of pavement structure.

5. Soak the sample for four days, with water available at the top and bottom.
6. Take a final swell-gauge reading. Remove the sample and mold from the water. Allow the sample to drain for 15 minutes.
7. Place it in the compression machine and seat the penetration piston on the surface with a 44.5-N (10-lb) load.
8. Apply a load on the piston, then record the total load when the piston has penetrated the soil to a depth of 2.54 mm (0.1 in.).

Results:

Swell-gauge readings: Initial   Final
Load required to cause 2.54-mm (0.1-in) penetration: (A)

Calculations:

Percentage swell = amount of swell/height of sample $\times$ 100 =

[Height = 116.4 mm (4.584 in.)]

CBR value = [load (A)/standard load] $\times$ 100 =

[Standard load at 2.54-mm (0.10-in.) penetration is 13.3 kN (3000 lb)]

5–6.2 *Compressive Strength of Cement-Modified Granular Materials*

Purpose: To measure compressive strength of a granular material that has been strengthened with portland cement.

Theory: When granular materials do not meet base course specifications or when extra-high strength is required, they are often strengthened by adding portland cement. Cylinders are cast with various cement contents. The design cement content is chosen on the basis of the least amount that produces the required compressive strength.

Apparatus: molds: standard 100-mm (4-in) compaction molds [or molds 70 mm (2.8 in) in diameter and 140 mm (5.6 in) high]
compression machine

Procedure:

1. Obtain the optimum moisture content from a standard compaction test, using the estimated cement content.
2. Compact the sample in a standard compaction mold with the standard effort. Level the surface.
3. Extract the sample from the mold and place it in a plastic bag for curing.
4. At the specified age—usually seven days—remove the sample and place it in a compression machine. Load until failure.

Results:

$$\text{Load at failure} \underline{\hspace{3cm}} \text{kN or lb}$$

Calculations:

$$\text{Compressive strength = load/area =} \underline{\hspace{2.5cm}} \text{kPa or lb/in}^2$$

5–7 PROBLEMS

5–1. What are the main layers in a pavement structure?

5–2. What are the two main types of pavement structure? How do they differ?

5–3. In a CBR test on a subgrade soil, a load of 1.73 kN (390 lb) was required to penetrate the sample the standard distance of 2.54 mm. What was the CBR value? Describe this soil as a subgrade. (Refer to table 5–3.)

5–4. Following are the test results on a subgrade soil for a proposed road: passing 75 μm (No. 200)–63%; w_L = 31; I_P = 19. Find the group index value and suggest a pavement structure based on the group index. Traffic is 4000 vehicles a day, 15% of which are trucks.

5–5. A pavement structure is proposed for a highway with a DTN of 2000. The average CBR of the subgrade is 8. Design the pavement structure according to the Asphalt Institute's charts for (a) full depth asphalt, and (b) asphalt with a high-quality base course.

5–6. A pavement is required to carry 350 equivalent 80 kN (18 000-lb) axle loads per day. The subgrade soil has a CBR of 3.5. The regional factor is 4, as frost heaving has occurred in the past in this area. The road is to be surfaced by 50 mm (2 in) of asphalt concrete, with 200 mm (8 in) of cement-treated base [strength: 3500 kPa (510 psi)] and a sand-gravel subbase. Find the depth of the subbase required, using the AASHTO method.

5—7. A local residential street is to be built on an ML subgrade which has a history of severe frost heaving. Design a pavement structure using the *Handbook of Highway Engineering* method for (a) a rigid pavement and (b) a flexible pavement.

5—8. Two types of subgrades are especially dangerous for highways. What are they, and why do they present problems?

5—9. A 10-m (33-ft) highway embankment is to be constructed with a fine sand material (A-3). What compaction requirements would you recommend?

5—10. What conditions are required for frost heaving to occur? What soils are most susceptible, and why?

5—11. Classify soils A through G in fig. 5—11 according to U.S. Army Corps of Engineers frost susceptibility criteria.

5—12. What is the subbase layer in (a) a flexible pavement, and (b) a rigid pavement?

5—13. Minimum and maximum allowable percentages are often specified for fines in highway bases. Why? What properties are affected by each limit and how?

5—14. What is meant by pumping? How does it happen?

5—15. Why is hardness important for base-course aggregate? What test is used to measure this property? What are the AASHTO requirements for hardness in base-course aggregates?

5—16. What are the AASHTO and ASTM limits for the liquid limit and the plasticity index for material to be used as base courses? Why are these limits specified?

5—17. The grain size distribution for a subgrade soil is:

Passing		
2.0 mm	100%
300 μm	85%
150 μm	60%
75 μm	42%
40 μm	15%

Three subbase materials (A, B, and C) are being considered for use over this subgrade. The D_{15} values for these three materials are 2.1 mm, 310 μm, and 80 μm, respectively. Check these subbases for protection against possible intrusion of subgrade material.

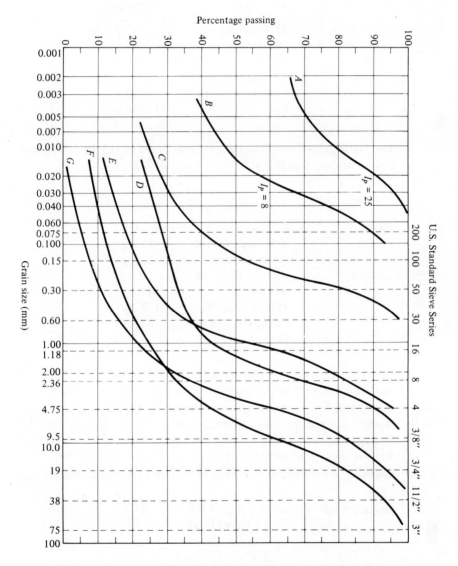

FIG. 5-11. Grain size distribution curves for various subgrade soils.

207

5—18. A granular sample is proposed for highway base course construction. Test results are:

Sieve	Retained
25 mm (1-in)	0 g
19 mm (3/4-in)	3011 g
12.5 mm (1/2-in)	3895 g
9.5 mm (3/8-in)	4113 g
4.75 mm (No. 4)	6349 g
Pan	11 943 g

The fine aggregate from the pan was split down to 513.4 g and washed. Loss in washing was 29.9 g. Results of fine sieving were:

Sieve	Retained
4.75 mm (No. 4)	0 g
2.00 mm (No. 10)	80.6 g
1.18 mm (No. 16)	122.0 g
600 mm (No. 30)	52.3 g
425 μm (No. 40)	35.1 g
150 μm (No. 100)	101.4 g
75 μm (No. 200)	83.7 g
Pan	7.3 g

Calculate the gradation. Plot the grain size distribution curve. Check for acceptance as base course material according to ASTM and AASHTO specifications.

5—19. Material specifications for a granular base course may include gradation limits, soundness, petrographic requirements, and percentage crushed. Why are these specified? (What property is affected, and how?)

5—20. Following are the results of a series of tests on a granular material:

Gradation test—Passing	19 mm (3/4 in)	100%
	9.5 mm (3/8 in)	71%
	4.75 mm (No. 4)	62%
	2.00 mm (No. 10)	47%
	600 μm (No. 30)	30%
	425 μm (No. 40)	16%
	300 μm (No. 50)	14%
	75 μm (No. 200)	12%

Abrasion test—Percent loss—38%

Atterberg limits—w_L = 22

$- I_P$ = 5

Fractured faces test—375 g fractured out of a total sample tested of 610 g.

(a) Evaluate this material according to the AASHTO specifications given in Standard M147. Include acceptable gradings and comments on all other quality requirements.

(b) Repeat, using ASTM Standard D2940.

5—21. Obtain the specifications used for granular base-course material by your local state or provincial highway department. Check the aggregate test results given in problem 5—20 against these specifications. What other tests would be required by your local authority?

5—22. What is soil cement used for in a pavement? What are the advantages of using soil cement? What tests are used to arrive at the percentage of cement required in a soil cement design?

5—23. A pavement is to be constructed with 150 mm (6 in) of soil-cement base over an area of 325 m by 162 m (1070 ft by 530 ft). The soil cement design mix is:

$$\rho_D = 1890 \text{ kg/m}^3 \ (118 \text{ lb/ft}^3)$$

$$w = 8.5\%$$

$$cc = 6.2\%$$

The soil for the soil cement is a beach sand trucked to a plant mixer at the site. The moisture content of the sand is 3.0%. The mixer produces 2000 kg (4400 lb) of soil cement per batch. Find

a. Mix proportions: mass of dry soil, cement, and water per m^3.

b. Field mix: mass of beach sand, cement, and extra water per m^3.

c. Number of m^3 of mix produced per batch. (*Note:* 2000 kg (4400 lb) is based on total density.)

d. Amount of beach sand, cement, and extra water required per batch.

e. Total number of m^3 (yd^3) of mix required to pave the area.

f. Total amount of cement required.

g. Total amount of beach sand required in m^3 (yd^3).

[*Note:* Beach sand has a loose density of 1700 kg/m^3 (106 lb/ft^3).]

5—24. Briefly describe soil types that can be best stabilized by (a) lime, (b) portland cement, and (c) asphalt.

5—25. A cement-modified granular material fails in a compression test at a total load of 26.7 kN (6000 lb). It was cast in a standard compaction mold. Find the compressive strength in kPa and lb/in^2.

5—26. A series of durability tests is conducted on soil-cement samples using a fine sand (A-3). Results are as follows:

Cement Content (%)	Percent Loss	
	Wet-dry Tests	Freeze-thaw Tests
7.0	16.2	19.5
8.0	14.8	16.6
9.0	13.5	14.3
10.0	13.0	11.8
11.0	12.7	10.2

What is the cement content that should be specified for this material?

5—27. Results of compressive strength tests (at 7 days) on three briquettes of cement-modified granular material, made with 8% cement are:

$$\text{Sample 1} \quad 1580 \text{ kPa (229 psi)}$$
$$2 \quad 1910 \text{ kPa (277 psi)}$$
$$3 \quad 1800 \text{ kPa (261 psi)}$$

The granular material is a medium fine sand with 17% smaller than 0.05 mm. Would this mixture meet the suggested requirements for cement-modified granular material? If not, what changes in the mix should be made for further tests?

5—28. What properties are important in geotextiles being used between aggregate bases and organic subgrades? Why?

5—29. A soft, saturated, uniform, fine sand is to be covered with a geotextile before the base course is placed. For the fine sand, $D_{85} = 400 \ \mu m$, and $D_{10} = 170 \ \mu m$. What properties should be specified regarding the required properties of the geotextile fabric? (Use Hazen's formula to estimate the permeability of the soil.)

5—30. What type of failures do geotextiles prevent in ditches and along shores? How? Where are the geotextile fabrics placed for this purpose?

6

Asphalt
Pavement Surfaces

Asphalt or bituminous materials combined with aggregates are the most common pavement surfaces in use today. They are used on all types of roadway —from multiple layers of asphalt concrete on the highest class of highway to thin, dust-control layers on seldom-used roads.

Portland cement concrete is the other main type of surface course, but it is seldom used in low-class roads. Properties, mix design, and quality control of portland cement concretes are covered in chapter 7.

6–1 ASPHALT PAVING MATERIALS

Bituminous materials (or *bitumens*) are hydrocarbons which are soluble in carbon disulphate. They are usually fairly hard at normal temperatures; when heated, they soften and flow. When mixed with aggregates in their fluid state, they solidify and bind the aggregates together, forming a pavement surface.

6–1.1 Bitumens that have been used in paving include:

1. *Native asphalts*. Obtained from asphalt lakes in Trinidad and other Caribbean areas, these were used in some of the earliest pavements in North America.
2. *Rock asphalts*. These are rock deposits containing bituminous materials which have been used for road surfaces in localities where they occur.

211

3. *Tars.* Tars are bituminous materials obtained from the distillation of coal.
4. *Petroleum asphalts.* These are products of the distillation of crude oil. These asphalts are by far the most common bituminous paving materials in use today, and are the only type discussed in this book.

6–1.2 Grades of asphalt materials and temperatures at which they are used depend to a great extent on their viscosity.

The viscosity of asphalt varies greatly with temperature, ranging from a solid to a fairly thin liquid. Viscosity-temperature relationships are extremely important in the design and use of these materials.

Viscosity decreases (that is, material becomes more fluid) as temperature increases. A very viscous fluid is very "thick."

 – Absolute (or dynamic) viscosity is measured in Pa · s (SI units) and poises (traditional units). [1 poise = 0.1 Pa · s.] Kinematic viscosity is measured in cm^2/s (SI units) and stokes or centistokes (traditional units) [1 stoke = 100 centistokes = 1 cm^2/s]. Since kinematic viscosity equals absolute viscosity divided by density (about 1 g/cm^3 for asphalts), the absolute viscosity and the kinematic viscosity have approximately the same numerical value when expressed in poises and stokes.

Viscosity has often been measured in the Saybolt Furol apparatus as the number of seconds it takes for a specified volume to flow. Approximate conversion is one second, Saybolt Furol (SSF) = two centistokes = 0.02 stokes.

Asphalt cements were originally graded according to *penetration value.* This is an empirical test in which the amount the needle penetrates a prepared asphalt sample in five seconds is measured in tenths of a millimeter under standard conditions. For example, if the needle penetrated 9.8 mm–or 98 tenths of a mm–the penetration value would be 98.

Figure 6–1 gives approximate relationships between various types of viscosity measurements.

Figure 6–2 illustrates typical temperature-viscosity relationships for asphalts. The AC 10 is harder than AC 5, and therefore to reduce the viscosity to, for example, 1.5 cm^2/s (150 centistokes), asphalt A must be heated to 167°C, and asphalt B only to 163°C.

Asphalt C, as illustrated, has a different temperature-viscosity relationship, as it probably came from a different crude oil field or was produced with a different refining process.

Plant temperatures for mixing asphalt paving materials are usually specified in terms of viscosity, for this indicates how fluid the material is and how well it will coat the aggregates without overheating. Temperature limits corresponding to viscosities of 1.5 to 3.0 cm^2/s (150 to 300 centistokes) are sometimes used. For asphalt A in fig. 6–2, this would require that the plant be operated at temperatures between 153°C and 167°C.

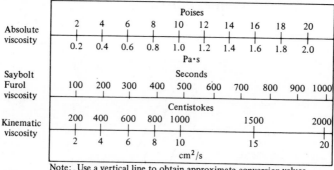

Note: Use a vertical line to obtain approximate conversion values.

FIG. 6—1. Approximate relationship of various units for measuring the viscosity of asphalts. (Modified with permission of The Asphalt Institute.)

The minimum temperature for spraying (as in pavement seal coats) is often specified as that corresponding to a viscosity of 2.0 cm^2/s (200 centistokes). For asphalt B in fig. 6-2, this would be 156°C.

6—1.3 Figure 6-3 shows a petroleum asphalt flow chart. The major paving products are:

— asphalt cements
— liquid asphalts
— asphalt emulsions

Asphalt cements are the primary asphalt products produced by the distillation of crude oil. They are produced in various viscosity grades, the most common being AC 2.5, AC 5, AC 10, AC 20, and AC 40. These roughly correspond to penetration grades 200-300, 120-150, 85-100, 60-70, and 40-50, respectively. The viscosity grades indicate the viscosity in hundreds of poises ± 20% measured at 60°C (140°F). For example, AC 2.5 has a viscosity of 250 poises ± 50. AC 40 has a viscosity of 4000 poises ± 800.

Liquid asphalts (or *cutback asphalts*) are asphalt cements mixed with a solvent to reduce their viscosity and, thus, make them easier to use at ordinary temperatures. They are commonly heated (if required) and then sprayed on aggregates. Upon evaporation of the solvent, they cure or harden and cement the aggregate particles together.

Types and grades are based on the type of solvent, which governs viscosity and the rates of evaporation and curing. The RC (rapid curing) types use gasoline as a solvent, and therefore cure rapidly. The MC (medium curing) types use kerosene. The SC (slow curing) types use diesel fuel, or they may be produced

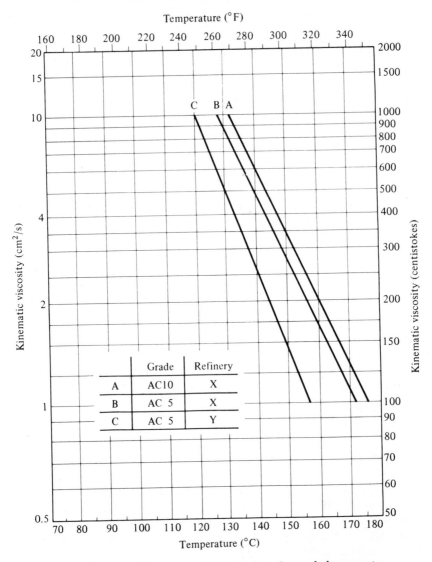

FIG. 6—2.   Temperature-viscosity chart for asphalt cements.

directly from the refinery during distillation. Solvent contents are commonly from 15% to 40% of the total. Grades of liquid asphalts are governed by viscosity, and are shown in fig. 6—4.

Asphalt emulsions are mixtures of asphalt cement and water. As these components do not mix themselves, an emulsifying agent (usually a type of soap) must be added. The emulsifying unit breaks up the asphalt cement and disperses it, in the form of very fine droplets, in the water carrier. When used, the emul-

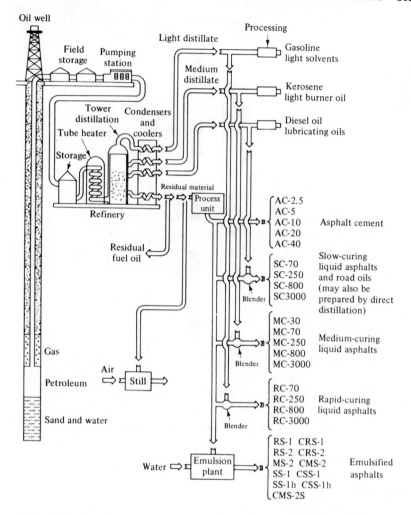

FIG. 6-3. Petroleum asphalt flow chart. (Modified with permission
of The Asphalt Institute.)

sion sets as the water evaporates. The emulsion usually contains 55%-75% asphalt
cement and up to 3% emulsifying agent, with the balance being water.

Two general types of emulsified asphalts are produced, depending on the
type of emulsifier used: *cationic* emulsions, in which the asphalt particles have a
positive charge; and *anionic*, in which they have a negative charge. Anionic emulsions adhere better to aggregate particles which have positive surface charges
(e.g., silica). Cationic emulsions also work better with wet aggregates and in
colder weather.

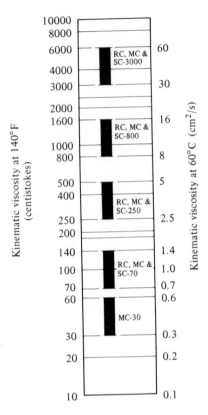

FIG. 6–4.   Liquid asphalt grades.

An emulsion's rate of hardening (or "breaking") depends on the amount and type of agent used. There are three grades of the two types of asphalts (C indicates cationic types):

1. Rapid setting (RS or CRS).
2. Medium setting (MS or CMS).
3. Slow setting (SS or CSS).

6–1.4   Asphalt cements are made up of three types of hydrocarbons: asphaltenes, resins, and oils.

Asphaltenes are large, hard, powdery, black materials. The asphaltenes are surrounded by resins, which contribute ductility and adhesiveness to asphalt. These particles are suspended in an oily material, which allows asphalt to flow. The consistency or hardness of asphalt cement depends on the relative proportions of these three materials in its composition.

Aging or hardening of asphalts is due to the evaporation and oxidation of the lighter oily constituents during mixing at high temperatures and to the oxidation of the oils to resins and resins to asphaltenes when used over a period of years. Design of asphalt mixtures must take into account these possible effects on the useful life of a pavement.

6–1.5   Quality control tests for asphalt materials include the following:

1. *Viscosity.* Many methods of measuring viscosity have been used, as indicated above in the discussion on viscosity:
— Absolute viscosity is measured by the vacuum capillary viscometer (ASTM D2171)
— Kinematic viscosity is measured by the kinematic viscometer (ASTM D2170)
— Viscosity in seconds, Saybolt Furol is measured in the Saybolt Furol apparatus (ASTM E102)
— Penetration values, measuring depth of penetration of a standard needle into asphalt cement, are obtained from the penetration apparatus (ASTM D5) (see fig. 6–5)

2. *Ductility (ASTM D113).* An asphalt sample is cast in a mold consisting of two jaws, then placed in a water bath. One jaw is moved from the other at a standard rate; the distance it moves before the thread between the two breaks is the ductility in centimeters (fig. 6–6).

3. *Thin Film Oven Test (ASTM D1754).* Asphalt paving materials in use are found as extremely thin layers joining aggregate particles together. The properties of the mix—especially durability—depend to a great extent on the properties of a thin film of asphalt. In this test, a thin sample is heated in an oven for a period of time, and the properties of the sample afterward are obtained as an indication of the rate of aging or hardening of the asphalt.

4. *Solubility (ASTM D2042).* With this test the purity of the asphalt can be checked.

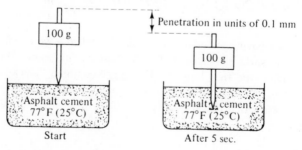

FIG. 6–5.   Penetration of asphalt cement. (Reprinted with permission of The Asphalt Institute.)

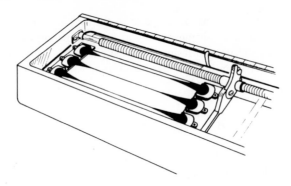

FIG. 6—6. Ductility test. (Reprinted with permission of The Asphalt Institute.)

5. *Flashpoint (ASTM D92)*. This test determines the temperature to which asphalt materials may safely be heated.

6—1.6 Asphalt cement grades and specifications have usually been based on penetration grades. Many agencies are currently switching to the more scientific viscosity grading system. ASTM specifications based on these systems are given in tables 6—1 and 6—2. In table 6—2, three specifications are available. Usually the requirements of table 1 govern, unless otherwise indicated by the purchaser. Authorities that specify asphalt cements based on properties of the residue from the thin-film oven test use grades AR–1000 to AR–16000 and the requirements given in the third table of table 6—2.

Specifications for liquid asphalts and asphalt emulsions are included in ASTM and AASHTO standards, as well as in many of the Asphalt Institute's booklets.

6—2 ASPHALT CONCRETE—PROPERTIES

The main asphalt paving material in use today is *asphalt concrete*. This is a high-quality pavement surface composed of asphalt cement and aggregates, hot-mixed in an asphalt plant and then hot-laid. This is the common "black-top" or "hot mix" or "asphalt" used on most roads (except rural and secondary roads) and most paved parking lots. Lower-quality pavement surfaces are discussed in subsequent sections.

6—2.1 Asphalt concrete consists of asphalt cement, aggregates, and air. However, some of the asphalt cement seeps into voids in the aggregate particles, and

Table 6-1

SPECIFICATIONS FOR PENETRATION GRADED ASPHALT CEMENTS (ASTM D946)*

| | Penetration Grade | | | | | | | | | |
| | 40-50 | | 60-70 | | 85-100 | | 120-150 | | 200-300 | |
	Min	Max	Min	Max	Min	Max	Min	Max	Min	Max
Penetration at 77°F (25°C) 100 g, 5 s	40	50	60	70	85	100	120	150	200	300
Flash point, °F (Cleveland open cup)	450	...	450	...	450	...	425	...	350	...
Ductility at 77°F (25°C) 5 cm/min, cm	100	...	100	...	100	...	100	...	100	...
Retained penetration after thin-film oven test, %	55+	...	52+	...	47+	...	42+	...	37+	...
Ductility at 77°F (25°C) 5 cm/min, cm after thin-film oven test	...	...	50	...	75	...	100	...	100	...
Solubility in trichloroethylene, %	99.0	...	99.0	...	99.0	...	99.0	...	99.0	...

* Reprinted by the permission of the American Society for Testing and Materials, 1916 Race Street, Philadelphia, PA 19103, Copyright.

Table 6–2

SPECIFICATIONS FOR VISCOSITY GRADED ASPHALT CEMENTS (ASTM STANDARD D3381)*

TABLE 1 Requirements for Asphalt Cement Viscosity Graded at 140°F (60°C)

Note—Grading based on original asphalt.

Test	Viscosity Grade				
	AC-2.5	AC-5	AC-10	AC-20	AC-40
Viscosity, 140°F (60°C), P	250 ± 50	500 ± 100	1000 ± 200	2000 ± 400	4000 ± 800
Viscosity, 275°F (135°C), min. cSt	80	110	150	210	300
Penetration, 77°F (25°C), 100 g, 5 s, min.	200	120	70	40	20
Flash point, Cleveland open cup, min. °F (°C)	325 (163)	350 (177)	425 (219)	450 (232)	450 (232)
Solubility in trichloroethylene, min. %	99.0	99.0	99.0	99.0	99.0
Tests on residue from thin-film oven test:					
Viscosity, 140°F (60°C), max. P	1250	2500	5000	10,000	20,000
Ductility, 77°F (25°C), 5 cm/min, min. cm	100ᵃ	100ᵃ	50	20	10

ᵃ If ductility is less than 100, material will be accepted if ductility at 60°F (15.5°C) is 100 minimum at a rate of 5 cm/min.

Table 6-2 continued

TABLE 2 Requirements for Asphalt Cement Viscosity Graded at 140°F (60°C)

Note—Grading based on original asphalt.

Test	Viscosity Grade				
	AC-2.5	AC-5	AC-10	AC-20	AC-40
Viscosity, 140°F (60°C), P	250 ± 50	500 ± 100	1000 ± 200	2000 ± 400	4000 ± 800
Viscosity, 275°F (135°C), min. cSt	125	175	250	300	400
Penetration, 77°F (25°C), 100 g, 5 s, min.	220	140	80	60	40
Flash point, Cleveland open cup, min. °F (°C)	325 (163)	350 (177)	425 (219)	450 (232)	450 (232)
Solubility in trichloroethylene, min. %	99.0	99.0	99.0	99.0	99.0
Tests on residue from thin-film oven test:					
Viscosity, 140°F (60°C), max. P	1250	2500	5000	10,000	20,00
Ductility 77°F (25°C), 5 cm/min, min. cm	100[a]	100[a]	75	50	25

[a] If ductility is less than 100, material will be accepted if ductility at 60°F (15.5°C) is 100 minimum at a rate of 5 cm/min.

Table 6-2 continued

TABLE 3 Requirements for Asphalt Cement Viscosity Graded at 140°F (60°C)

Note—Grading based on residue from rolling thin-film oven test.

Tests on Residue from Rolling Thin-Film Oven Test[a]	Viscosity Grade					
	AR-1000	AR-2000	AR-4000	AR-8000	AR-16000
Viscosity, 140°F (60°C), P	1000 ± 250	2000 ± 500	4000 ± 1000	8000 ± 2000	16000 ± 4000
Viscosity, 275°F (135°C), min. cSt	140	200	275	400	550
Penetration, 77°F (25°C), 100 g, 5 s, min.	65	40	25	20	20
% of original penetration, 77°F (25°C), min.	...	40	45	50	52
Ductility, 77°F (25°C), 5 cm/min, min. cm	100[b]	100[b]	75	75	75
Tests on original asphalt: Flash Point, Cleveland Open Cup, min. °F(°C)	400 (205)	425 (219)	440 (227)	450 (232)	460 (238)
Solubility in trichloroethylene, min. %	99.0	99.0	99.0	99.0	99.0

[a] Thin-film oven test may be used but the rolling thin-film oven test shall be the referee method.

[b] If ductility is less than 100, material will be accepted if ductility at 60°F(15.5°C) is 100 minimum at a rate of 5 cm/min.

* Reprinted by the permission of the American Society for Testing and Materials, 1916 Race Street, Philadelphia, PA 19103, Copyright.

222

therefore is not available to coat and bind aggregates together. This also leaves more air voids in the mixture than would be expected by calculating the total aggregate and asphalt volumes. Figure 6–7 shows the components of an asphalt concrete. Relative amounts of aggregate, asphalt, and air are important, as is discussed in the following section.

The amount of asphalt absorption is less than the water absorption for the same aggregates, usually by about one-half. However, it is important to include the volume of absorbed asphalt in calculations, since all volumes must be measured accurately. The amount of asphalt absorption can be found by measuring the relative density of a mixture of asphalt-coated aggregates, and comparing this with the value expected with no absorption. Procedures for a test to obtain this are included in this chapter's section on test methods.

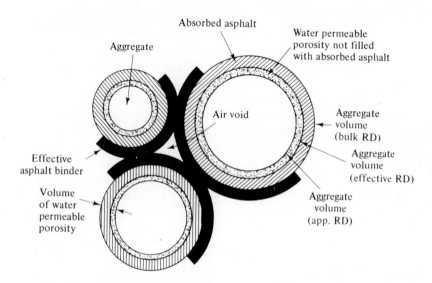

FIG. 6–7. Asphalt mixture showing net or effective asphalt, absorbed asphalt, and air voids.

The mass-volume relationships for asphalt (bituminous) concrete are illustrated in fig. 6–8.

The following relationships are usually calculated:

$$\text{Density } (\rho) = M/V \qquad (1–3)$$

$$\text{Asphalt content (AC)} = M_B/M \qquad (6–1)$$

$$\text{Asphalt absorption (Asp Abs)} = M_{BA}/M_G \qquad (6–2)$$

$$\text{Air voids (AV)} = V_A/V \qquad (6-3)$$

$$\text{Voids in mineral aggregate (VMA)} = (V_A + V_{BN})/V \qquad (6-4)$$

(See fig. 6–8.)

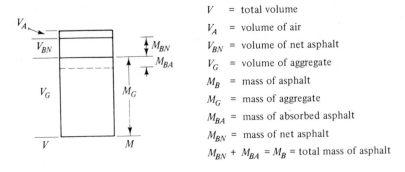

V = total volume
V_A = volume of air
V_{BN} = volume of net asphalt
V_G = volume of aggregate
M_B = mass of asphalt
M_G = mass of aggregate
M_{BA} = mass of absorbed asphalt
M_{BN} = mass of net asphalt
$M_{BN} + M_{BA} = M_B$ = total mass of asphalt

FIG. 6–8.   Mass-volume relationships for asphalt concrete.

Example 6–1: An asphalt concrete mix contains 2250 kg of aggregates and 150 kg of asphalt per m³. Asphalt absorption is 1.2%. The bulk relative density of the aggregates is 2.67; relative density of the asphalt, 1.05. Find the mass-volume relationships.

$M_G = 2250$ kg

$M_B = 150$ kg

$M = 2400$ kg

$M_{BA} = 0.012 \times 2250 = 27$ kg

$M_{BN} = 150 - 27 = 123$ kg

$V_G = 2250/(2.67 \times 1000) = 0.843$ m³

$V_{BN} = 123/(1.05 \times 1000) = 0.117$ m³

$V_A = 1 - (0.843 + 0.117) = 0.040$ m³

Therefore

$$AC = 150/2400 = 6.25\%$$

$$\rho = 2400/1 = 2400 \text{ kg/m}^3$$

$$AV = 0.040/1 = 4.0\%$$

$$VMA = (0.040 + 0.117)/1 = 15.7\%$$

Usually the density is measured; the asphalt content, asphalt absorption, and relative density values are known; and the voids relationships must be calculated.

Example 6–2:

Given:
$$\rho = 2440 \text{ kg/m}^3$$
$$AC = 5.8\%$$
$$\text{Asp Abs} = 0.8\%$$
$$RD_B \text{ (aggregates)} = 2.67$$
$$RD \text{ (asphalt)} = 1.03$$

Find AV and VMA.

$$M_B = 0.058 \times 2440 = 142 \text{ kg}$$
$$M_G = 2440 - 142 = 2298 \text{ kg}$$
$$M_{BA} = 0.008 \times 2298 = 18 \text{ kg}$$
$$M_{BN} = 142 - 18 = 124 \text{ kg}$$
$$V_G = 2298/(2.67 \times 1000) = 0.861 \text{ m}^3$$
$$V_{BN} = 124/(1.03 \times 1000) = 0.120 \text{ m}^3$$
$$V_A = 1 - (0.861 + 0.120) = 0.019 \text{ m}^3$$

Therefore

$$AV = 0.019/1 = 1.9\%$$
$$VMA = (0.019 + 0.120)/1 = 13.9\%$$

Some organizations calculate the asphalt content as a percentage of the mass of aggregates, not as a percentage of the total mass. Calculations required for this approach are illustrated in the following example.

Example 6–3:

Given:
$$\text{Density} = 148.5 \text{ lb/ft}^3$$
$$AC = 6.5\% \text{ (as percentage of mass of aggregates)}$$
$$\text{Asp Abs} = 1.1\%$$

Bulk relative density (aggregates) = 2.61

Relative density (asphalt cement) = 1.04

Find AV and VMA.

$$M_G = 148.5/(1 + 0.065) = 139.44 \text{ lb}$$

M

$$M_B = 148.5 - 139.44 = 9.06 \text{ lb}$$

$$M_{BA} = 0.011 \times 139.44 = 1.53 \text{ lb}$$

$$M_{BN} = 9.06 - 1.53 = 7.53 \text{ lb}$$

$$V_G = 139.44/(2.61 \times 62.4) = 0.856 \text{ ft}^3$$

$$V_{BN} = 7.53/(1.04 \times 62.4) = 0.116 \text{ ft}^3$$

$$V_A = 1 - (0.856 + 0.116) = 0.028 \text{ ft}^3$$

Therefore

$$AV = 0.028/1 = 2.8\%$$

$$VMA = (0.028 + 0.116)/1 = 14.4\%$$

6–2.2 Asphalt concrete surfaces must provide smooth, skid-resistant riding surfaces. They must be strong enough to carry the imposed loads without rutting. They must maintain these properties for the design life. Since they distribute loads by deflecting slightly with each load application, they must be flexible. These requirements lead to the following required properties for asphalt concrete mixes:

— strength
— flexibility
— durability
— skid resistance

Most specifications for asphalt concrete take into account the necessity of meeting these four requirements.

Strength must be sufficient to carry the load without shear occurring between particles. The structure must remain intact. The main contributor to strength is friction between the grains. A dense-graded mixture is best for high friction strength with a relatively low amount of binder. If the asphalt coating around the particles is too thick, the amount of friction between particles is reduced.

Flexibility is obviously very important, as these are flexible pavements. The asphalt concrete must be able to deflect slightly under each load without cracking. For this requirement, a more open-graded aggregate mixture is better, as is a higher asphalt cement content. These conditions allow more movement without cracking.

Durability measures the pavement's resistance to wear and aging. Aggregates should be hard and cubical to ensure the minimum breakdown during manufacture and during application of loads. Aggregates should also be sound, not susceptible to disintegration from cycles of freezing and thawing. Certain aggregates have a greater affinity for water than for asphalt cement. In these cases, water may replace the asphalt film on the aggregate particles, destroying the bond between particles.

The major causes of asphalt concrete aging are evaporation and oxidation of asphalt cement. During mixing at high temperatures, some of the lighter constituents of the asphalt evaporate, leaving a harder cement. After construction, air and water circulate through the material. These lead to oxidation of the asphalt, again removing the lighter constituents and leaving a hard, brittle material. Figure 6–9 illustrates the effects of evaporation and high rates of oxidation on asphalt cements, showing how either of these may reduce the pene-

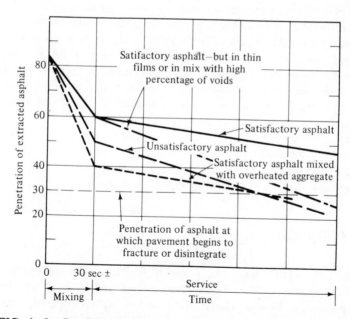

FIG. 6–9. Possible changes in penetration of asphalt paving mixtures during mixing and service. (Reprinted with permission from Clarkson H. Oglesby, *Highway Engineering*, 3rd ed. New York: Wiley, 1975.)

tration value of the cement to about 30, a level at which there is evidence that cracking will occur.

Cracking leads to rapid failure of the pavement, since it loses some of its load distribution properties and allows water into the surface and base, again lowering load-carrying capacity.

To control aging and hardening of the binder materials, the following are often specified:

1. Maximum temperature during mixing, to reduce evaporation.
2. Maximum percentage of air voids to reduce permeability and movement of air and water in the mixture, and therefore to reduce the rate of oxidation.
3. Minimum percentage VMA, to ensure that sufficient space is left for asphalt cement (which helps to ensure that the binder film around each particle is thick enough to remain ductile, not brittle).
4. The softest possible grade of asphalt cement for a project, softer grades being less likely to crack in cold weather.

Loss of *skid resistance* of asphalt concrete surfaces is mainly caused by *polishing* of the aggregates or *bleeding* of the cement. Surface courses usually have a lower maximum particle size in order to increase the number of small particles and therefore the number of projections at the surface for skid resistance. The aggregates should be hard and resistant to wear, and thus resistant to polishing. It has been found that limestone aggregates tend to polish in many cases. Bleeding occurs on hot days, when the cement tends to seep to the surface in mixtures with few voids. Specifications usually require a minimum air void content so that asphalt cement can be accommodated in the air void space as the pavement becomes denser under load.

Mix requirements to meet the above criteria are summarized in table 6–3. Obviously no one mixture is best for all these properties, and a compromise must be made in specifications to accommodate each property to the maximum extent possible without seriously affecting other properties.

6–3 AGGREGATES

Aggregates for asphalt concrete are usually classified as coarse aggregates, fine aggregates, or mineral filler. Types of coarse and fine aggregate have been discussed previously. Mineral filler is often used in asphalt concrete mixtures to supply the fines (smaller than 75 μm or No. 200 sizes). Fines are very important in producing a dense-graded, strong material. Many natural sands do not contain the amount or type of fines required. Limestone dust is the most common material used for mineral filler.

Table 6–3

PROPERTIES AND MIX REQUIREMENTS FOR
ASPHALT CONCRETE

Property	Asphalt Content	Aggregate Gradation	Air Voids	Aggregate Quality
Strength	Low	Dense	Low	Rough faces; crushed
Flexibility	High	Open	High	Coarser sizes; better
Durability	High	Dense	Low	Hard, cubical; resistant to freeze-thaw damage; does not strip
Skid Resistance	Low	—	High	High sand content; hard; resistant to polishing

6–3.1   Aggregates should be

1. *Well-graded*—dense, including mineral filler (if required) for strength.
2. *Hard*—for resistance to wear and to polishing due to traffic.
3. *Sound*—for resistance to breakdown due to freezing and thawing.
4. *Rough surfaced*—crushed rough surfaces give higher friction strength and a better surface for adhesion of the asphalt cement.
5. *Cubical*—thin, elongated aggregate particles break easily.
6. *Hydrophobic (or "water hating")*—some siliceous aggregates such as quartz are hydrophilic ("water liking"), meaning that they have a greater affinity for water than for asphalt, due to their surface charges. This may lead to stripping, as asphalt coating comes away from the particle in the presence of water.

   Measurement of these properties has been discussed previously, except for the test to measure susceptibility to stripping (ASTM D1075). In this test samples are immersed in water for specified periods. The resulting loss in strength is compared with control samples to indicate resistance to stripping.
   Typical specifications for aggregates for dense-graded asphalt concrete are shown in tables 6–4, 6–5, and 6–6.

Table 6—4

SPECIFICATIONS FOR MINERAL FILLER FOR ASPHALT CONCRETE (AASHTO M17)*

Mineral filler shall be graded within the following limits:

Sieve	Mass Percentage Passing
0.600 mm (No. 30)	100
0.300 mm (No. 50)	95–100
0.075 mm (No. 200)	70–100

The mineral filler shall be free from organic impurities and have a plasticity index not greater than 4.

* Reprinted by permission of the American Association of State Highway and Transportation Officials from *AASHTO Materials,* 12th edition, 1978.

Table 6—5

SPECIFICATIONS FOR FINE AGGREGATE FOR ASPHALT CONCRETE (AASHTO M29)

Grading Requirements:

Sieve Size	Amounts Finer than Each Laboratory Sieve (Square Openings) Mass Percent		
	Grading No. 1	Grading No. 2	Grading No. 3
9.5 mm (3/8 in.)	100	. . .	100
4.75 mm (No. 4)	95 to 100	100	80 to 100
2.36 mm (No. 8)	70 to 100	95 to 100	65 to 100
1.18 mm (No. 16)	40 to 80	85 to 100	40 to 80
0.600 mm (No. 30)	20 to 65	65 to 90	20 to 65
0.300 mm (No. 50)	7 to 40	30 to 60	7 to 40
0.150 mm (No. 100)	2 to 20	5 to 25	2 to 20
0.075 mm (No. 200)	0 to 10	0 to 5	0 to 10

Additional Requirements: Loss in soundness test shall be less than 15% when sodium sulphate is used, and 20% when magnesium sulphate is used.

* Reprinted by permission of the American Association of State Highway and Transportation Officials from *AASHTO Materials,* 12th edition, 1978.

Table 6–6

SPECIFICATIONS FOR COARSE AGGREGATE FOR ASPHALT CONCRETE (ASTM D692)

Grading Requirements:

Size No.	Nominal Size (Sieves with Square Openings)	Amounts Finer than Each Laboratory Sieve (Square Openings), Weight Percent									
		2½-in. (63-mm)	2-in. (50-mm)	1½-in. (37.5-mm)	1-in. (25.0-mm)	3/4-in. (19.0-mm)	1/2-in. (12.5-mm)	3/8-in. (9.5-mm)	No. 4 (4.75-mm)	No. 8 (2.36-mm)	No. 16 (1.18-mm)
3	2 to 1-in. (50 to 25.0-mm)	100	90 to 100	35 to 70	0 to 15	…	…	…	…	…	…
357	2-in. to No. 4 (50 to 4.75-mm)	100	95 to 100	…	35 to 70	…	10 to 30	…	0 to 5	…	…
4	1½ to 3/4-in. (37.5 to 19.0-mm)	…	100	90 to 100	20 to 55	0 to 15	…	0 to 5	…	…	…
467	1½-in. to No. 4 (37.5 to 4.75-mm)	…	100	95 to 100	…	35 to 70	…	10 to 30	0 to 5	…	…
5	1 to 1/2-in. (25.0 to 12.5-mm)	…	…	100	90 to 100	20 to 55	0 to 10	0 to 5	…	…	…
57	1-in. to No. 4 (25.0 to 4.75-mm)	…	…	100	95 to 100	…	25 to 60	…	0 to 10	0 to 5	…
6	3/4 to 3/8-in. (19.0 to 9.5-mm)	…	…	…	100	90 to 100	20 to 55	0 to 15	0 to 5	…	…

Table 6–6 (continued)

67	3/4-in. to No. 4 (19.0 to 4.75-mm)	...	100	90 to 100	...	20 to 55	0 to 10	0 to 5	...
68	3/4-in. to No. 8 (19.0 to 2.36-mm)	...	100	90 to 100	...	30 to 65	5 to 25	0 to 10	0 to 5
7	1/2-in. to No. 4 (12.5 to 4.75-mm)	...	...	100	90 to 100	40 to 70	0 to 15	0 to 5	...
78	1/2-in. to No. 8 (12.5 to 2.36-mm)	...	...	100	90 to 100	40 to 75	5 to 25	0 to 10	0 to 5
8	3/8-in. to No. 8 (9.5 to 2.36-mm)	...	...	...	100	85 to 100	10 to 30	0 to 10	0 to 5

Other Requirements:

Percent crushed: minimum of 40% by mass of particles retained on the No. 4 shall have at least one fractured face

Soundness: maximum loss in soundness test (magnesium sulphate)—18%

Abrasion: maximum loss in Los Angeles Abrasion Test:
—40% for surface courses
—50% for base courses

* Reprinted by permission of the American Society for Testing and Materials, 1916 Race Street, Philadelphia, PA 19103, Copyright.

6–4 ASPHALT CONCRETE MIX DESIGN

The design of an asphalt concrete mixture includes the selection of the best blend of aggregates and the optimum asphalt content to provide a material that meets the required specifications as economically as possible.

Mix design involves the following steps:

1. Selection of aggregate proportions to meet the specification requirements.
2. Conducting trial mixes at a range of asphalt contents and measuring the resulting physical properties of the samples.
3. Analyzing the results to obtain the optimum asphalt content and to determine if the specifications can be met.
4. Repeating with additional trial mixes using different aggregate blends, until a suitable design is found.

The two most common methods for making and evaluating trial mixes are the *Marshall method* and the *Hveem method*.

6–4.1 Specifications for the gradation of the blended aggregates in asphalt concrete vary considerably. A number of gradation specifications for various types of asphalt materials and for different types of pavements are given in ASTM Standard D3515.

Typical specifications for base and surface courses are given in table 6–7.

In the first step in a mix design, proportions of the proposed aggregates (including mineral filler) are chosen to produce a combined gradation close to the center of the specification limits using methods such as shown in section 4–6. Either the Marshall method or the Hveem method can then be used to make the first trial mix.

6–4.2 The *Marshall Method* consists of the following major steps:

1. Aggregates are blended in proportions that meet the specification's requirements.
2. The mixing and compacting temperatures for the asphalt cement being used are obtained from the temperature-viscosity graph. These temperatures are those required to produce viscosities of $1.7 \pm .2$ cm^2/s (170 ± 20 centistokes) for mixing and $2.8 \pm .3$ cm^2/s (280 ± 30 centistokes) for compacting.
3. A number of briquettes, 101.6 mm (4 in) in diameter and 60–65 mm (2½ in) high, are mixed using 1200 g of aggregates and asphalt cement at various percentages both above and below the expected optimum content. For surface courses with 12.5 mm (1/2-in.) aggregate, the expected optimum content

233

Table 6-7

TYPICAL GRADATION REQUIREMENTS
FOR ASPHALT CONCRETE

Sieve	Allowable Percent Passing	
	Base Courses	Surface Courses
38 mm (1½ in)	100	
25 mm (1 in)	90–100	
19 mm (3/4 in)	–	100
12.5 mm (½ in)	60–80	90–100
9.5 mm (3/8 in)	–	70–95
4.75 mm (No. 4)	25–55	45–70
1.18 mm (No. 16)	15–40	20–50
300 μm (No. 50)	4–20	5–25
75 μm (No. 200)	2–8	3–10
(Approx. asphalt content	4%–7%	5%–8%)

may be about 6.5%. Therefore briquettes would be made at 5.5%, 6.0%, 6.5%, 7.0%, and 7.5% asphalt cement.

4. Density of the briquettes is measured to allow calculation of the voids' properties.

5. Briquettes are heated to 60°C (140°F). *Stability* and *flow* values are obtained in a compression test in the Marshall apparatus to measure strength and flexibility. The stability is the maximum load that the briquettes can carry. The flow is the compression (measured in units of hundredths of an inch or in millimetres) that the sample undergoes between no load and maximum load in the compression test.

Photographs of the Marshall test apparatus are shown in figures 6–10 and 6–11.

Results of the Marshall test are plotted on graphs such as those shown in fig. 6–12–density, stability, flow, air voids, and VMA are plotted against asphalt content. These typical relationships can be observed:

1. Density initially increases with asphalt content, since the fluid lubricates grain movements. Eventually, however, a maximum density is reached. Then density decreases, since the lighter asphalt replaces some of the aggregate, shoving the particles apart.

2. Stability increases and decreases along with asphalt content on a curve similar to that for density, since the strength is mainly a function of friction between grains of aggregates and, therefore, of density.

FIG. 6–10. Marshall Compaction Apparatus–the Marshall model, with base plate and collar attached, is in the mold holder, with the Marshall hammer at the side.

FIG. 6–11. Marshall Stability Load Frame.

3. Flow increases along with asphalt content, since friction between particles decreases with thicker asphalt films.
4. The percentage of air voids decreases as asphalt content increases, since the asphalt tends to fill all the void spaces.
5. The percentage of voids in mineral aggregate is approximately opposite to the density curve, since the mass of aggregates is the main component of the total mass of the mix.

The optimum asphalt content is one that economically and safely satisfies all specification requirements.

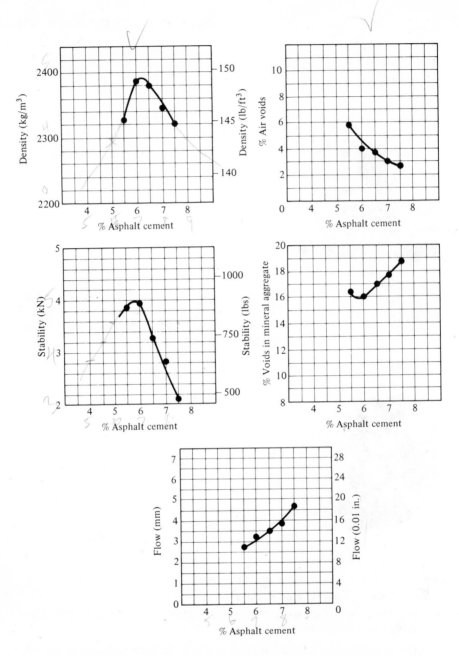

FIG. 6–12. Typical test results from a Marshall Mix design.

The Asphalt Institute's suggested specifications are shown in table 6–8. Evaluation of a trial mix is illustrated in example 6–4.

Table 6–8

**SUGGESTED CRITERIA FOR ASPHALT CONCRETE DESIGN
BY THE MARSHALL METHOD***

	Heavy Traffic		Medium Traffic		Light Traffic	
	Min	*Max*	*Min*	*Max*	*Min*	*Max*
Number of compaction blows, each end of specimen	75		50		35	
Stability (lb)	750	–	500	–	500	–
(N)	3,336		2,224		2,224	
Flow (units of 0.01 in.)	8	16	8	18	8	20
(mm)	2	4	2	4.5	2	5
% Air Voids						
Surfacing or Leveling	3	5	3	5	3	5
Base	3	8	3	8	3	8
% Voids in Mineral Aggregate						
Nominal Maximum Size of Aggregate						
No. 4 (4.75 mm)	18		18		18	
3/8-in. (9.5 mm)	16		16		16	
1/2-in. (12.5 mm)	15		15		15	
3/4-in. (19 mm)	14		14		14	
1-in. (25 mm)	13		13		13	
1½-in. (38 mm)	12		12		12	

* Adapted by permission of The Asphalt Institute.

Example 6–4: Results of a trial mix have been plotted in fig. 6–12. The mix is to meet the Asphalt Institute's requirements for a surface course subjected to medium traffic, with 12.5-mm (1/2-in) maximum sized aggregates. From graphs:

Asphalt content at maximum density	6.2%
Asphalt content at maximum stability	5.8%
Asphalt content at 4% air voids (the middle of the 3–5% allowed)	6.3%
Average asphalt content	6.1%

This mix meets all requirements:

Stability	4000 N (900 lb)
Flow	3 mm (12 units of 0.01 in)
AV	4.3%
VMA	16.0%

6–4.3 The following example illustrates the calculations required by the Marshall Method.

Example 6–5:

1. *Aggregate Blending.* A combination with 45% coarse aggregate (A), 50% fine aggregate (B), and 5% mineral filler (C) is arrived at by methods outlined in chapter 4 for blending aggregates. To ensure that no segregation occurs in measuring out samples for briquettes, aggregate A and aggregate B are each divided into two sizes. Of aggregate A, 58% passes 3/8-in.; of aggregate B, 38% passes No. 16. So to make a briquette containing 1200 g of aggregates, the following proportions are used:

Aggregate A: Larger than 3/8-in:	$0.42 \times 45 = 18.9\% \times 1200$ g $= 227$ g
(total 45%) Smaller than 3/8-in:	$0.58 \times 45 = 26.1\% \times 1200$ g $= 313$ g
Aggregate B: Larger than No. 16:	$0.62 \times 50 = 31.0\% \times 1200$ g $= 372$ g
(total 50%) Smaller than No. 16:	$0.38 \times 50 = 19.0\% \times 1200$ g $= 228$ g
Aggregate C:	$5.0\% \times 1200$ g $=\ \underline{\ \ 60}$ g
(total 5%)	Total 1200 g

2. Assuming that the temperature-viscosity relationship for the asphalt cement is represented by line *A* on Fig. 6–2, find:

Mixing temperature	(190–150) centistokes	161–167°C
Compacting temperature	(310–250) centistokes	152–156°C

3. The sample is mixed with an asphalt content as designed. (Assume 6.4%.)
4. The sample is compacted into a briquette with 35, 50, or 75 blows of the Marshall hammer on each face. (See table 6–8.)
5. Density determination:

Mass in air	1226.4 g
Mass submerged	721.9
Volume	504.5 cm^3
Density	1226.4/504.5 = 2.431 g/cm^3 = 2431 kg/m^3

6. Voids calculations:

| Density | 2431 kg/m^3 |
| AC | 6.4% |

Relative density of aggregates: Use a weighted average of the given relative densities for the three aggregates (an approximate relationship, but close enough for voids calculations), 2.68, 2.71 and 2.64, respectively.

Aggregate A: RD_B = 2.68 × 0.45 = 1.206

Aggregate B: RD_B = 2.71 × 0.50 = 1.355

Aggregate C: RD_B = 2.64 × 0.05 = 0.132

∴RD of aggregates =   2.693

Also given is:

| Asphalt absorption | 0.9% |
| Relative density of asphalt | 1.06 |

∴ taking a 1 m^3 volume:

$$M_B = 0.064 \times 2431 = 156 \text{ kg}$$

$$M_G = 2431 - 156 = 2275 \text{ kg}$$

$$M_{BA} = 0.009 \times 2275 = 20 \text{ kg}$$

$$M_{BN} = 156 - 20 = 136 \text{ kg}$$

$$V_G = 2275/(2.693 \times 1000) = 0.845 \text{ m}^3$$

$$V_{BN} = 136/(1.06 \times 1000) = 0.128 \text{ m}^3$$

$$V_A = 1 - (0.845 + 0.128) = 0.027 \text{ m}^3$$

$$AV = 2.7\%$$

$$VMA = 15.5\%$$

7. The sample is tested in the compression machine, the maximum load is recorded as 5780 N (1300 lb). The volume correction for a 2½-in. thickness of the sample (see the lab instruction sheet in section 6–8.2) is 1.04. The corrected stability is 1.04 × 5780 = 6010 N (1350 lb). The flow dial reads 29 (units of 0.01 inches) at the beginning of compression and 40 at the end. Therefore the flow is 11, or 2.8 mm.

6–4.4 The next most common method of mix design in North America is the Hveem method. The main steps in this design are

1. Obtain the estimated optimum asphalt content by the centrifuge kerosene equivalent method.
2. Prepare test briquettes at a range of asphalt contents above and below the estimated optimum.
3. Conduct stabilometer tests to obtain stabilometer values in the Hveem apparatus.
4. Conduct swell tests on two samples containing the estimated optimum asphalt content.

See the Asphalt Institute's "Mix Design Methods for Asphalt Concrete" for a detailed procedure. Table 6–9 gives suggested specifications.

6–4.5 A laboratory design includes the proportions of each aggregate, asphalt cement, and filler or other additives for the mixture, the temperature range for operation of the plant, and the quality required in the final product.

Table 6–9
SUGGESTED CRITERIA FOR ASPHALT CONCRETE DESIGNED BY THE HVEEM METHOD

	Heavy Traffic		Medium Traffic		Light Traffic	
	Min	Max	Min	Max	Min	Max
Stabilometer value	37	—	35	—	30	—
Swell (in.)	—	.030	—	.030	—	.030
(mm)	—	0.762	—	0.762	—	0.762
% Voids	4	—	4	—	4	—

* Reprinted by permission of The Asphalt Institute.

For example, following is a typical mix design.

Composition	Percent
Asphalt cement	6.4
Aggregates*	
coarse (crushed rock)	50
fine (sand)	35
fine (screenings)	15

Temperature	°C
Plant minimum	148
Plant maximum	161
Paver minimum	128

Mix properties

Density	2380 kg/m^3 (148.5 lb/ft^2)
Marshall stability	5300 N (1190 lb)
Marshall flow	3 mm (12 units of 0.10 in)
VMA	15.5%
AV	3.4%

* Gradation of each aggregate and the combined aggregate
   graduation curve are also included with the mixed design data.

6–5 PRODUCTION AND QUALITY CONTROL

After an acceptable mix design has been determined, the mix must be produced
and placed in an acceptable manner. Quality control checks and tests must be
followed.

6–5.1 Traditional asphalt plants have five components, as shown in fig. 6–13.
These are:

1. Cold aggregate storage bins.
2. Dryer (with dust collection facilities).
3. Gradation unit or screens.
4. Hot storage bins.
5. Pug mill for mixing.

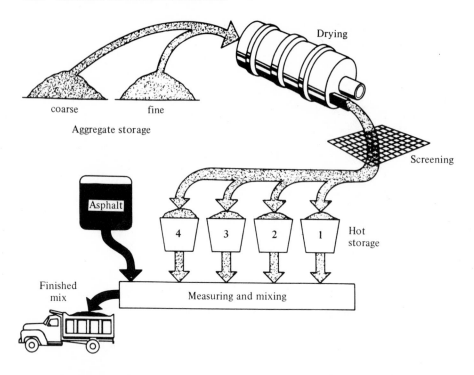

FIG. 6–13. Schematic diagram of a traditional asphalt plant. (Reprinted with permission of The Asphalt Institute.)

A surge bin containing the mixed asphalt is often included as part of the plant to store the material until it can be discharged into trucks. Storage silos are also used to store large volumes, and can keep a mixture for a number of days without seriously altering its properties. Heat must be supplied to maintain the temperature of asphalt cement, and oxygen must be removed from the silo and replaced with an inert gas to prevent oxidation of the asphalt cement.

These asphalt plants use two methods of measuring materials into the pug mill for mixing.

In older, "continuous" plants, aggregates are fed continually into the pug mill, sprayed with asphalt cement, mixed, and emptied into trucks. The amount of aggregate from each hot bin is controlled by the gate opening, and the asphalt cement is pumped in at a rate which is calibrated according to the rate that the aggregates enter the pug mill.

In more modern "batch" plants, the asphalt cement and aggregates from each hot bin are weighed into the pug mill to form a batch. The batch is mixed and discharged. The two plants are illustrated in fig. 6–14, and photographs of a batch plant are shown in fig. 6–15.

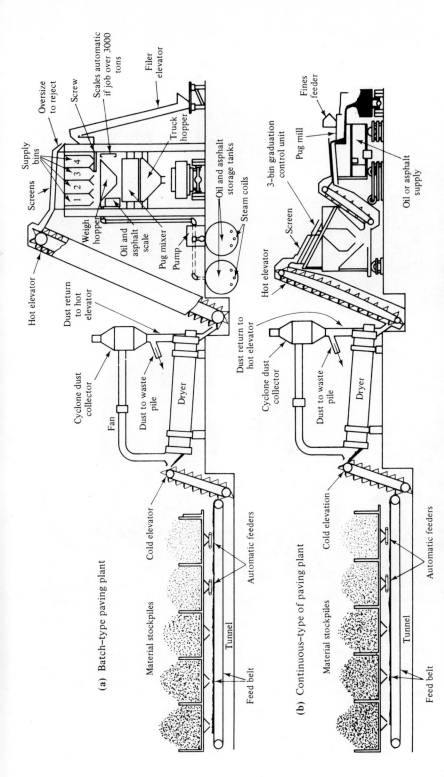

FIG. 6—14. Batch and continuous types of asphalt plants. (Reprinted with permission of The Asphalt Institute.)

(a) Batch-type paving plant

Oversize to reject

Screw

Scales automatic if job over 3000 tons

Filer elevator

Supply bins

Screens

1 2 3 4

Truck hopper

Weigh hopper

Oil and asphalt scale

Oil and asphalt storage tanks

Pug mixer

Pump

Steam coils

Hot elevator

Dust return to hot elevator

Cyclone dust collector

Fan

Dust to waste pile

Dryer

Cold elevator

Automatic feeders

Material stockpiles

Tunnel

Feed belt

(b) Continuous-type of paving plant

Fines feeder

3-bin graduation control unit

Pug mill

Oil or asphalt supply

Screen

Hot elevator

Dust return to hot elevator

Cyclone dust collector

Dust to waste pile

Dryer

Cold elevation

Automatic feeders

Material stockpiles

Tunnel

Feed belt

243

(a)

(b)

FIG. 6–15. Asphalt plant: (a) cold feed bins, containing (from left) 19-mm (3/4-in) crushed rock, 9.5-mm (3/8-in) crushed rock, screenings, and sand (storage silos for mixed material can be seen in the background); (b) drying and mixing plant—(1) conveyor from cold bins, (2) dryer, (3) screening unit and hot bins, (4) mixer (pugmill), (5) conveyor to storage silos, (6) conveyor for special additives (in this case asbestos fibers, supplied in bags, are being added to the mix in the pugmill.) (Courtesy of D. Cruppi and Sons.)

6–5.2 Quality control at the plant includes checks on the following aspects of production:

— quality and proportions of cold feed aggregates
— temperature
— proportions from hot bins and amount of asphalt cement, fillers, etc. added
— quality of resulting mixture.

Daily samples are taken of the material in each cold bin for gradation tests, to ensure that the type of material being used has not changed significantly from that used in the design. The gates from each cold bin must be adjusted to result in the designed proportions. In some plants the amount of material from each of the cold bins is controlled by scales.

Temperatures of aggregates as they leave the dryer, of the asphalt cement, and of the final mixture are usually recorded automatically and continuously. Aggregates from the dryer are usually about 10°-20°C higher than that of the final mixture. Samples of aggregates from the dryer may have to be tested periodically to ensure that all the water has been removed.

As the aggregates have been combined in the dryer and separated again by size in the gradation unit above the hot bins, the amount required from each hot bin must be established. Calculations involved are shown in the following example.

Example 6–6: The mix design for an asphalt project is:

1. Aggregates: 50% Aggregate A—Coarse aggregate
                35% Aggregate B—Fine aggregate
                15% Aggregate C—Screenings
(The combined grain size distribution is shown on the grain size distribution curve in fig. 6–16.) Asphalt content—6.2%.
2. Mix properties:

Stability	5300 N (1190 lb)
Density	2380 kg/m^3 (148.5 lb/ft^3)
VMA	15.5%
AV	3.8%
Flow	3 mm (12 units)

3. Mixing temperatures: 135–145°C

The plant is a batch type, producing 1360 kg (3000 lb) per batch. Material is stored in three hot bins before mixing. Tests were made on samples from the three bins, with results as follows:

Coarse Bin

Passing 25 mm (1-in.) 100%
12.5 mm (1/2-in.) 73%
9.5 mm (3/8-in.) 18%
2.36 mm (No. 8) 1%

Medium Bin

Passing 9.5 mm (3/8-in.) 100%
4.75 mm (No. 4) 75%
2.36 mm (No. 8) 33%
1.18 mm (No. 16) 11%

Fine Bin

Passing 2.36 mm (No. 8) 100%
1.18 mm (No. 16) 73%
600 μm (No. 30) 52%
300 μm (No. 50) 33%

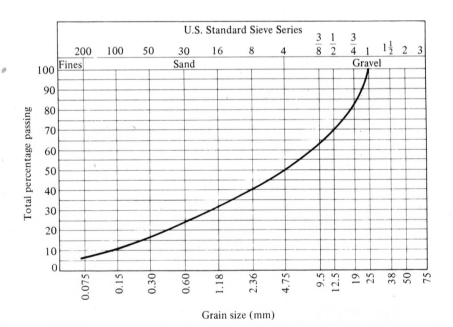

FIG. 6–16. Grain size distribution curve for an asphalt concrete mixture.

Calibration of the plant: (Only SI units used in calculations).

Cold bins: Gate openings are adjusted so that relative amounts from the three bins are 50:35:15 for aggregates A, B, and C respectively.

The temperature of the aggregates from the dryers and asphalt cement is adjusted to the required amount.

Hot bins:

$$\text{Total mass required per batch} = 1360 \text{ kg}$$

$$\text{Asphalt cement } (6.2\% \times 1360 = \quad 84 \text{ kg}$$

$$\text{Aggregates } (100 - 6.2)\% \times 1360 = 1276 \text{ kg}$$

From an examination of the results of tests on the contents of the hot bin, the coarse bin is mainly larger than 9.5 mm; the medium bin, 9.5 to 2.36 mm; the fine bin, smaller than 2.36 mm.

From the required grain-size distribution curve, the mix required is:

$$\begin{array}{lc}
\text{Passing 25 mm} & 100\% \\
9.5 \text{ mm} & 63\% \\
2.36 \text{ mm} & 41\%
\end{array}$$

or

$$37\% \text{ of } 1276 = 472 \text{ kg larger than 9.5 mm}$$

$$22\% \text{ of } 1276 = 281 \text{ kg between 9.5 and 2.36 mm}$$

$$41\% \text{ of } 1276 = 523 \text{ kg smaller than 2.36 mm}$$

Since 82% of the material in the coarse bin is larger than 9.5 mm, required from the coarse bin:

$$\frac{472}{0.82} = 576 \text{ kg}$$

of which
$$\begin{array}{l}
82\% \text{ is larger than 9.5 mm} = 472 \text{ kg} \\
17\% \text{ is between 9.5 and 2.36 mm} = \quad 98 \text{ kg} \\
1\% \text{ is smaller than 2.36 mm} = \quad 6 \text{ kg}
\end{array}$$

Required from the medium bin:

$$281 \text{ kg } (9.5 \text{ to } 2.36 \text{ mm}) - 98 \text{ kg (coarse bin)} = 183 \text{ kg}$$

Since 67% of medium bin is between 9.5 and 2.36 mm, required from medium bin:

$$\frac{183}{0.67} = 273 \text{ kg}$$

of which

67% is between 9.5 and 2.36 mm = 183 kg

33% is smaller than 2.36 mm =   90 kg

Required from the fine bin: 523 kg smaller than 2.36 mm − 6 kg (coarse bin) and 90 kg (medium bin) = 427 kg.
Therefore, the final quantities (kg) required are:

		Larger Than 9.5 mm	9.5 to 2.36 mm	Smaller Than 2.36 mm
Coarse bin	576	472	98	6
Medium bin	273		183	90
Fine bin	427			427
Totals	1276	472	281	523

Quantities required from each hot bin are measured into the pugmill by scales in batch plants and by adjustment of hot-bin gates in continuous plants. Also the quantity of asphalt cement and mineral filler or other additives, such as asbestos fiber, if required, must be established.

Samples of the mixture as it comes from the pugmill are taken for the preparation of briquettes to measure the strength, flow, and voids properties and to conduct extraction and gradation tests to check mixture composition. In the extraction test, the asphalt is dissolved and removed from the mix in a centrifuge; gradation of the remaining aggregates is determined by sieving. (The test procedure is given in section 6−8.4.)

6−5.3 Drum mixing plants are more economical for the production of asphalt cement and are becoming more common in the industry.

The steps in production are effectively reduced from four (cold feed, drying, hot gradation, mixing) to two (cold feed and drying-mixing) as follows:

1. Cold feed bins store the aggregates. Scales control the amount of material from each bin to ensure that the correct proportions are maintained.

2. Aggregate enters the drum. Drums in these plants have the burner located at the aggregate entrance end rather than at the exit end as in other plants. As the aggregate enters and proceeds through the drum, the following occurs:

   a. Surface and free moisture is evaporated.

   b. The temperature of the aggregate is raised to 75-80°C (170-180°F).

   c. Asphalt cement is introduced. The aggregate is now at 80-90°C (180-200°F). The moisture that is driven off causes the asphalt to foam, trapping dust particles and engulfing the aggregates.

   d. Further mixing occurs and the temperature is increased to the specified level.

Quality control is simplified as the proportions are continually controlled automatically in most plants. The amount of aggregate entering the drum is recorded as the material flows over the belt conveyor load cell. The amount of asphalt cement being introduced into the drum is automatically adjusted to correspond to the amount of aggregate. Moisture content of the aggregates must be measured to allow a correction factor to be applied to the aggregate quantities entering the drum. Temperature of the mix leaving the drum is measured and usually the burner is automatically adjusted to achieve the required mix temperature, which is lower than that required in traditional plants.

Surge bins or silos are usually used with drum plants to allow for various rates of transportation of the mix and to help control possible segregation.

Drum plants usually have higher production rates (up to 5000 tonnes or 5500 tons per day) than traditional plants.

A modern drum plant is illustrated in fig. 6-17.

6–5.4 After transportation to the construction site, the asphalt concrete must be laid, finished by the paver, and rolled. The main aspects of quality control, in addition to grade and thickness requirements, are:

1. *Temperature*. There must be a minimum temperature at the paver to ensure that the concrete can be compacted and that aggregate particles can be worked into the requisite dense, strong structure.

2. *Compaction*. The layer must be compacted to meet specifications, usually expressed as a percentage of the laboratory density.

Density is obtained from a sample removed from the pavement or by a nuclear densometer.

Asphalt paving equipment and compaction equipment are shown in figs. 6–18 and 6–19 respectively.

FIG. 6–17. Drum mixing plant: (1) aggregate bins; (2) scales control amount of material flowing from each bin; (3) screen to remove over-size particles; (4) total amount of aggregate measured by a weight idler; (5) radiation zone where aggregate is heated and excess water is driven off; (6) convection/coating zone where asphalt is introduced and coating occurs; (7) elevator for mixed material; (8) surge silo. (Courtesy of the Barber-Greene Company.)

6–6 ASPHALT CONCRETE RECYCLING

Reuse of materials in asphalt pavements is a growing trend due to the large increases in costs of petroleum products and, to a lesser extent, the growing shortage of quality aggregates in some areas.

FIG. 6—18. Asphalt pavers. Note the sensor (1) touching the grade reference beam; and (2) controlling the finished grade of the surface. (Courtesy of the Barber-Green Company.)

Recycling involves removal and crushing existing pavements, heating, addition of new asphalt cement or other rejuvenating additives, and new aggregates, and relaying.

6—6.1 Two main methods are used in reclaiming the existing pavement.

The conventional method involves breaking up the pavement with a bull-

(a)   (b)

FIG. 6-19. Compaction of asphalt concrete pavement: (a) pneumatic-tired roller for initial "breakdown" compaction; (b) tandem smooth-wheeled roller for final compaction and smoothing, shown finishing a joint between lanes.

dozer equipped with ripper teeth or with a backhoe, loading and transporting to the plant where the material is crushed to the required size, usually 38 mm (1½ in) or less.

In the cold milling method, the pavement is removed in small particles. The milling machine consists of a large rotating drum with teeth that cut into the pavement and remove chunks. The size of the particles obtained and the depth of cut in the pavement can be adjusted.

The milling method is generally found to be superior. Material can be removed to any depth and reduced to required sizes. The depth removed can be varied to leave a surface that meets specifications for profile grade and transverse slope, thus correcting any rutting or other surface irregularities in the pavement.

6-6.2 In a traditional batch or continuous plant usually up to about 50% reclaimed material can be used. It is usually added unheated to one of the hot bins. The new aggregates are superheated and when mixed with the reclaimed aggregate in the pug mill raise its temperature to soften old asphalt cement and allow mixing of the old and new asphalt cement and coating or recoating of the aggregate particles.

With drum plants, up to 70% reclaimed to 30% new material can be mixed. The new aggregate enters the drum at the burner end as usual and is rapidly heated. Reclaimed material is added usually about half way along the drum and is heated by the new aggregate and by the hot gases from the burner. The new asphalt is added further along the drum and mixing completed. A drum plant equipped to produce recycled asphalt is illustrated in fig. 6-20.

FIG. 6–20. Drum recycling plant: (1) bins for new aggregate; (2) bin for reclaimed material; (3) tank for asphalt cement or other rejuvenating agents; (4) amount of new aggregate and of reclaimed material measured by weight idlers; (5) radiation zone in the drum where new aggregate is super-heated and water evaporated; (6) center feed inlet where reclaimed material enters the drum mixer; (7) convection/coating zone where new asphalt and recycling agents are introduced and coating and mixing occur; (8) surge silo; (9) dust control and reclaiming equipment. (Courtesy of the Barber-Greene Company.)

Surge silos are important for all types of plants. The temperature of reclaimed pavement materials continues to rise in the silo, softening the old asphalt cement for more uniform mixing of the old and new asphalt.

6–6.3 Mix design methods are the same as for new asphalt concrete, using the Marshall or Hveem methods.

About 1.5%–4.0% new asphalt is required, a significant saving from the 4%–8% range usually contained in asphalt concrete from virgin material.

The new asphalt is one grade (or more) softer than the original, for example, AC 5 (Pen. 120-150) instead of AC 10 (Pen. 85-100). The reclaimed asphalt is much harder than it was originally due to aging. The resultant mixture should be close to the original design in grade of asphalt cement.

Other chemical rejuvenating agents or other types of asphalts such as emulsions or cut-backs can also be used in recycling projects.

6–6.4 Recycling (or rejuvenation) of asphalt pavement surfaces can be completed in place with equipment such as heater-planers, heater-scarifiers or hot milling machines. The reconstruction process is completed in one pass. Heat softens the asphalt in the old pavement. Surface imperfections, such as ruts, or surface layers of aged, cracked pavement are removed up to a depth of about 25 mm (1 in) by planing, scarifying or milling. New aggregate materials and asphalt cement or other rejuvenating agents are often added to the reclaimed material during mixing. A new surface is then laid and rolled.

Equipment is also available to break up thicker layers of asphalt pavement, pulverize the material, and mix and lay a new pavement surface in place.

6–7 OTHER ASPHALT SURFACES

Other asphalt pavememts can be classified as either surface treatments of various types or low-quality mixtures.

6–7.1 *Surface treatments* are low-class surfaces in which an asphalt material is sprayed on the surface, then covered with aggregates. The aggregates, either chips (12- or 6-mm; 1/2- or 1/4-in) or sand, are then rolled. A surface of about 12 mm (1/2 in) results. Additional layers may be placed on this to build up a pavement structure.

Tables 6–10 and 6–11 give recommended types and quantities of surface treatment materials.

Seal coats are surface treatments whose primary purpose is to seal cracks in existing asphalt surfaces. Dust is also controlled by this type of surface. Sand is usually the aggregate used in these applications.

Table 6–10

QUANTITIES OF MATERIALS FOR SURFACE TREATMENTS
(ASTM STANDARD RECOMMENDED PRACTICE D1369)*

(a) Traditional U.S. Units

Surface Treatment			Aggregate		Bituminous Material[a]
Type	Application	Size No.[b]	Nominal Size (Square Openings)	Typical Rate of Application (ft^3/yd^2)	Typical Rate of Application (gal/yd^2)
Single	initial	5	1 in. to 1/2 in.	0.50	0.42
		6	3/4 in. to 3/8 in.	0.36	0.37
		7	1/2 in. to No. 4	0.23	0.23
		8	3/8 in. to No. 8	0.17	0.19
		9	No. 4 to No. 16	0.11	0.13
Double	initial	5	1 in. to 1/2 in.	0.50	0.42
	second	7	1/2 in. to No. 4	0.25	0.26
Double	initial	6	3/4 in. to 3/8 in.	0.36	0.37
	second	8	3/8 in. to No. 8	0.18	0.20
Triple	initial	5	1 in. to 1/2 in.	0.50	0.42
	second	7	1/2 in. to No. 4	0.25	0.26
	third	9	No. 4 to No. 16	0.13	0.14
Triple	initial	6	3/4 in. to 3/8 in.	0.36	0.37
	second	8	3/8 in. to No. 8	0.18	0.20
	third	9	No. 4 to No. 16	0.13	0.14

(b) Metric Units

Surface Treatment			Aggregate		Bituminous Material[a]
Type	Application	Size No.[b]	Nominal Size (Square Openings, mm)	Typical Rate of Application (m^3/m^2)	Typical Rate of Application $(liter/m^2)$
Single	initial	5	25.0 to 12.5	0.017	1.90
		6	19.0 to 9.5	0.012	1.68
		7	12.5 to 4.75	0.008	1.04
		8	9.5 to 2.36	0.006	0.86
		9	4.75 to 1.18	0.004	0.59

Table 6–10 continued

Double	initial	5	25.0 to 12.5	0.017	1.90
	second	7	12.5 to 4.75	0.008	1.18
Double	initial	6	19.0 to 9.5	0.012	1.68
	second	8	9.5 to 2.36	0.006	0.91
Triple	initial	5	25.0 to 12.5	0.017	1.90
	second	7	12.5 to 4.75	0.008	1.18
	third	9	4.75 to 1.18	0.004	0.63
Triple	initial	6	19.0 to 9.5	0.012	1.68
	second	8	9.5 to 2.36	0.006	0.91
	third	9	4.75 to 1.18	0.004	0.63

* Reprinted by the permission of the American Society for Testing and Materials, 1916 Race Street, Philadelphia, PA 19103, Copyright.

[a] Experience has shown that these quantities should be increased slightly (5 to 10 percent) when the bituminous material to be used was manufactured for application with little or no heating.

[b] According to ASTM Specification D 448, Standard Sizes of Coarse Aggregate for Highway Construction, which appears in the *Annual Book of ASTM Standards,* Part 15. (Gradation limits also show in Table 6.6.)

Prime coats, also known as *tack coats*, consist of asphalt sprayed on either an existing base or a portland cement or asphalt surface to help bond new construction.

6–7.2 *Low-quality mixes* include those mixed either in a plant (cold-laid) or on the road. Types, designs, and quality controls vary greatly.

6–7.3 Table 6–12 lists recommended uses of asphalt materials for many types of paving. Table 6–13 gives recommended application temperatures for asphalts.

6–8 TEST PROCEDURES

The following procedures are outlined below. Detailed standard procedures may be found in the appropriate standards, which are listed in the Appendix.

6–8.1 Marshall Briquettes
6–8.2 Marshall Stability
6–8.3 Asphalt Absorption
6–8.4 Asphalt Extraction Test

Table 6–11

RECOMMENDED GRADES OF ASPHALT MATERIALS FOR
SURFACE TREATMENTS (ASTM STANDARD
RECOMMENDED PRACTICE D1369)*

Nominal Size (Square Openings), in. (mm)	Size No.**	Hot Weather (80 F +) (26.7 C +)		Cool Weather (50 to 80 F) (10 to 26.7 C)	
1 to 1/2 (25.0 to 12.5)	5	MC	3000	MC	3000
		RC	3000	RC	3000
		RS	2	RS	2
		CRS	2	CRS	1, 2
		120 to 150 Pen.		120 to 150 Pen.	
3/4 to 3/8 (19.0 to 9.5)	6	MC	3000	MC	800
		RC	3000	RC	800
		RS	2	RS	2
		CRS	1, 2	CRS	1, 2
		120 to 150 Pen.			
1/2 to No. 4 (12.5 to 4.75)	7	MC	3000	MC	800
		RC	800, 3000	RC	250, 800
		RS	2	RS	2
		CRS	1, 2	CRS	1, 2
		200 to 300 Pen.			
3/8 to No. 8 (9.5 to 2.36)	8	RC	250, 800	RC	250, 800
		RS	1, 2	RS	1, 2
		CRS	1, 2	CRS	1, 2
No. 4 to No. 16 (4.75 to 1.18)	9	RC	250, 800	RC	250, 800
		RS	1, 2	RS	1, 2
		CRS	1, 2	CRS	1, 2

* Reprinted by the permission of the American Society for Testing and Materials, 1916 Race Street, Philadelphia, PA 19103, Copyright.

** See Table 6.6 for aggregate gradation limits.

6–8.1 *Marshall Briquettes*

Purpose: To make asphalt briquettes for (1) determination of voids' properties, and (2) subsequent testing for strength and flexibility in the Marshall stability apparatus.

Theory: Aggregates are blended to meet specifications; a series of briquettes are made at a range of asphalt contents. The briquettes are 101.6 mm (4 in) in

Table 6-12

USES FOR ASPHALT MATERIALS IN PAVING

Type of Construction	Viscosity Graded -Original					Viscosity Graded -Residue					Penetration Graded				
	AC-40	AC-20	AC-10	AC-5	AC-2.5	AR-160	AR-80	AR-40	AR-20	AR-10	40-50	60-70	85-100	120-150	200-300
ASPHALT-AGGREGATE MIXTURES															
Asphalt Concrete and Hot-Laid Plant Mix															
Pavement Base and Surfaces															
Highways		X	X	X	X	X	X	X	X	X		X	X	X	X
Airports	X	X					X	X				X	X		
Parking Areas	X	X				X	X	X				X	X		
Driveways	X	X					X	X	X			X	X		
Curbs	X	X	X			X	X				X	X	X		
Industrial Floors	X	X				X	X				X	X			
Blocks	X										X				
Groins	X	X									X	X			
Dam Facings	X	X									X	X			
Canal and Reservoir Linings	X	X									X	X			
Cold-Laid Plant Mix															
Pavement Base and Surfaces															
Open-Graded Aggregate															
Well-Graded Aggregate															
Patching, Immediate Use															
Patching, Stockpile															
Mixed-in-Place (Road Mix)															
Pavement Base and Surfaces															
Open-Graded Aggregate															
Well-Graded Aggregate															
Sand		X	X											X	X
Sandy Soil		X	X											X	X
Patching, Immediate Use															
Patching, Stockpile															
ASPHALT-AGGREGATE APPLICATIONS															
Surface Treatments															
Single Surface Treatment		X	X											X	X
Multiple Surface Treatment		X	X											X	X
Aggregate Seal		X	X					X						X	X
Sand Seal															
Slurry Seal															
Penetration Macadam															
Pavement Bases															
Large Voids		X												X	
Small Voids			X												X
ASPHALT APPLICATIONS															
Surface Treatment															
Fog Seal															
Prime Coat, Open Surfaces															
Prime Coat, Tight Surfaces															
Tack Coat															
Dust Laying															
Mulch															
Crack Filling															
Asphalt Pavements															
Portland Cement Concrete Pavements	X										X				

* Reprinted by permission of The Asphalt Institute.

	Rapid Curing (RC)				Medium Curing (MC)					Slow Curing (SC)				Emulsified — Anionic							Emulsified — Cationic					
	70	250	800	3000	30	70	250	800	3000	70	250	800	3000	RS-1	RS-2	MS-1	MS-2	MS-2h	SS-1	SS-1h	CRS-1	CRS-2	CMS-2	CMS-2h	CSS-1	CSS-1h
								X						X	X	X					X	X				
		X					X	X	X	X	X	X							X	X	X				X	X
	X	X					X	X	X		X	X							X	X					X	X
							X	X		X	X															
	X	X	X					X	X	X	X			X	X	X					X	X				
	X						X	X		X	X								X	X	X				X	X
	X	X				X	X	X											X	X					X	X
	X	X	X				X	X											X	X	X				X	X
	X	X					X	X	X		X	X							X	X					X	X
							X	X		X	X															
	X	X	X					X	X					X	X						X	X				
	X	X	X						X					X	X						X	X				
	X	X	X					X	X					X	X						X	X				
	X						X	X						X	X						X	X				
																			X	X					X	X
		X	X											X	X						X	X				
	X													X	X						X	X				
																			X	X					X	X
	X	X				X	X																			
	X			X	X														X							
	X													X					X	X	X				X	X
						X													X	X					X	X
																			X	X					X	X

Table 6–13

TYPICAL TEMPERATURES FOR THE USE OF ASPHALT

| Type and Grade of Asphalt | Pugmill Mixture Temperatures (°C) | | Spraying Temperatures (°C) | |
	Dense-Graded Mixes	Open-Graded Mixes	Road Mixes	Surface Treatments
Asphalt Cements				
AC-2.5	115–140	80–120	–	130+
AC-5	120–145	80–120	–	140+
AC-10	120–155	80–120	–	140+
AC-20	130–165	80–120	–	145+
AC-40	130–170	80–120	–	150+
AR-10	105–135	80–120	–	135+
AR-20	135–165	80–120	–	140+
AR-40	135–165	80–120	–	145+
AR-80	135–165	80–120	–	145+
AR-160	150–175	80–120	–	–
200–300 pen.	115–150	80–120	–	130+
120–150 pen.	120–155	80–120	–	130+
85–100 pen.	120–165	80–120	–	140+
60–70 pen.	130–170	80–120	–	145+
40–50 pen.	130–175	80–120	–	150+
Cutback Asphalts (RC, MC, SC)				
30 (MC only)	–	–	–	30+
70	–	–	20+	50+
250	55–80	–	40+	75+
800	75–100	–	55+	95+
3000	80–115	–	–	110+
Emulsified Asphalts				
RS-1	–		–	20–60
RS-2	–		–	50–80
MS-1	10–70		20–70	–
MS-2	10–70		20–70	–
MS-2h	10–70		20–70	–
SS-1	10–70		20–70	–
SS-1h	10–70		20–70	–
CRS-1	–		–	20–60
CRS-2	–		–	50–80
CMS-2	10–70		20–70	–
CMS-2h	10–70		20–70	–
CSS-1	10–70		20–70	–
CSS-1h	10–70		20–70	–

* Reprinted by permission of The Asphalt Institute.

Type and Grade of Asphalt	Pugmill Mixture Temperatures (°F)		Spraying Temperatures (°F)	
	Dense-Graded Mixes	Open-Graded Mixes	Road Mixes	Surface Treatments
Asphalt Cements				
AC-2.5	235–280	180–250	—	270+
AC-5	250–295	180–250	—	280+
AC-10	250–315	180–250	—	280+
AC-20	265–330	180–250	—	295+
AC-40	270–340	180–250	—	300+
AR-10	225–275	180–250	—	275+
AR-20	275–325	180–250	—	285+
AR-40	275–325	180–250	—	290+
AR-80	275–325	180–250	—	295+
AR-160	300–350	180–250	—	—
200–300 pen.	235–305	180–250	—	265+
120–150 pen.	245–310	180–250	—	270+
85–100 pen.	250–325	180–250	—	280+
60–70 pen.	265–335	180–250	—	295+
40–50 pen.	270–350	180–250	—	300+
Cutback Asphalts (RC, MC, SC)				
30 (MC only)	—	—	—	85+
70	—	—	65+	120+
250	135–175	—	105+	165+
800	165–210	—	135+	200+
3000	180–240		—	230+
Emulsified Asphalts				
RS-1	—		—	70–140
RS-2	—		—	125–175
MS-1	50–160		70–160	—
MS-2	50–160		70–160	—
MS-2h	50–160		70–160	—
SS-1	50–160		70–160	—
SS-1h	50–160		70–160	—
CRS-1	—		—	70–140
CRS-2	—		—	125–175
CMS-2	50–160		70–160	—
CMS-2h	50–160		70–160	—
CSS-1	50–160		70–160	—
CSS-1h	50–160		70–160	—

diameter and 60–65 mm (2½ in) high. After hardening, their mass is obtained in air. Then they are submerged to find their density and, hence, voids' properties.

Apparatus: Marshall mold with base and collar
             Marshall compaction pedestal
             Marshall hammer
             balance
             oven

Procedure:

1. Measure out 1200 g of aggregates, blended in the desired proportions. Heat the aggregates in the oven to the mixing temperature.
2. Add asphalt cement at the mixing temperature to give the desired asphalt content.
3. Mix the materials in a heated pan with heated mixing tools.
4. Return the mixture to the oven and reheat it to the compacting temperature.
5. Place the mixture in a heated Marshall mold with a collar and base. Spade the mixture around the sides of the mold. Place filter papers under the sample and on top of the sample.
6. Place the mold in the Marshall compaction pedestal.
7. Compact the material with 50 blows of the hammer (or as specified), invert the sample, and compact the other face with the same number of blows.
8. After compaction, invert the mold. With the collar on the bottom, remove the base and extract the sample by pushing it out with the extractor.
9. Allow the sample to stand for a few hours to cool.
10. Obtain the sample's mass in air and submerged.

Results:

Mass of aggregates in mixing pan	1200 g	
Mass of asphalt cement added	_____	g
Asphalt content	_____	%
Mixing temperature – aggregates	_____	°C
– asphalt	_____	°C
Compacting temperature	_____	°C
Number of blows with hammer per face	_____	
Mass of briquette	_____	g
Mass submerged	_____	g
Volume of briquette	_____	cm³
Density	_____	g/cm³

Calculations: Using given or known values for the bulk relative density of aggregates and asphalt cement and for asphalt absorption, calculate the air voids and voids in the mineral aggregate.

6–8.2 Marshall Stability

Purpose: To measure the strength and flexibility of asphalt mixtures using briquettes prepared by the Marshall method.

Theory: Strength and flexibility are two important properties required of asphalt concrete for use in pavements. In the Marshall test, the prepared briquettes are heated to 60°C (140°F). This temperature represents the weakest condition for an asphalt pavement in use. This test measures the maximum load that the sample can carry when tested in compression, with the load applied to the circumference of the sample. This is the Marshall "stability." Also measured (at the same time) is the change in diameter of the sample in the direction of load application between the start of loading and the time of maximum load. This change in diameter indicates the brittleness or flexibility of the mixture. Stability values are corrected to allow for variation from the standard 2½-in thickness.

Apparatus: Marshall loading yoke
flow meter
compression machine
water bath

Procedure:

1. Place the briquettes in a water bath at 60° ± 1°C (140° ± 1.8°F) for 30–40 minutes.
2. Place one briquette in the loading yoke, add the top part of the yoke, place the flow meter over one of the posts, and adjust it to read zero.
3. Apply a load at a rate of 50 mm (2 in) per minute until the maximum load reading is obtained.
4. Record the maximum load reading. At the same instant, obtain the flow as recorded on the flow meter. (*Note:* A stopwatch can be used to measure the time from start of loading to maximum load. Using a rate of loading of 50 mm (2 in)/minute, the flow in millimetres or units of 1/100 in can be calculated.)

Results: Record the maximum load as uncorrected stability in N or lb. Record the flow in units of 1/100 in. or in mm.

Calculations: Correct the recorded stability values by multiplying by a factor dependent on the volume of the briquette.

MARSHALL STABILITY CORRECTION FACTORS

Volume (cm^3)	Factor
457–470	1.19
471–482	1.14
483–495	1.09
496–508	1.04
509–522	1.00
523–535	0.96
536–546	0.93
547–559	0.89
560–573	0.86

6–8.3 Asphalt Absorption

Purpose: To measure asphalt absorption by aggregates in an asphalt mixture.

Theory: Aggregates are porous materials. The aggregate particle absorbs water in surface pores. Asphalt, when mixed with aggregate at a high temperature, is also very fluid and soaks into pores in aggregate particles. Asphalt that is in pores is not available to coat and cement aggregate particles. The asphalt remaining takes up less of the void space between aggregate particles, thus creating more air voids.

Apparatus: pycnometer (1 liter)
vacuum pump
balances

Procedure:

1. Add 800–1000 g of dry mix to the pycnometer. Obtain the mass.
2. Add water to cover the sample.
3. Apply a vacuum to remove the air.
4. Fill the pycnometer approximately to the calibration mark. Place it in a water bath to bring the temperature to approximately 20°C. Check to be sure that the water level is at the calibration mark. Obtain the mass.

Results:

Mass of pycnometer filled with water at 20°C (from laboratory standards)	_____	g (A)
Mass of mix + pycnometer	_____	g (B)
Mass of pycnometer	_____	g (C)
Mass of vacuumed mix + water + pycnometer	_____	g (D)

Calculations:

Data on materials and mixture

Asphalt content of mix	_____	%
Relative density (bulk) of aggregates	_____	(RD_G)
Relative density of asphalt	_____	(RD_B)
Mass of asphalt mix (B – C)	_____	g (M)
Mass of pycnometer + water	_____	g (A)
Total (A + M)	_____	g (E)
Mass of pycnometer + mix + water	_____	g (D)
Volume of mixture (E – D)	_____	cm^3 (V)
Maximum relative density of mixture (M/V)	_____	
Mass of asphalt (% asphalt $\times$ M)	_____	g (M_B)
Mass of aggregates ($M - M_B$)	_____	g (M_G)
Volume of asphalt (M_B/RD_B)	_____	cm^3 (V_B)
Volume of aggregates (M_G/RD_G)	_____	cm^3 (V_G)
Total ($V_B + V_G$)	_____	cm^3 (F)
Volume absorbed asphalt (F – V)	_____	cm^3 (V_{BA})
Mass absorbed asphalt ($V_{BA} \times RD_B$)	_____	g (M_{BA})
Absorption (M_{BA}/M_G)	_____	%

6–8.4 Asphalt Extraction Test

Purpose: To check the asphalt content and aggregate gradation of an asphalt mixture.

Theory: Quality control of asphalt pavement construction may require the checking of asphalt content and aggregate gradation. In this test, a sample is placed in a centrifuge. A solvent is added to dissolve the asphalt cement. The solvent and asphalt are thrown out of the sample, through a filter, by centrifugal force.

Apparatus: centrifuge with bowl for sample
              balances
              sieves
              trichlorethane

Note: Trichlorethane is a toxic substance, and should be used only under a fume hood or in a well-ventilated area with a surface exhaust system.

Procedure:

1. Obtain a sample of about 1000 g of asphalt mix.
2. Break the sample down to small pieces with a fork. The sample can be heated to about 115°C (240°F).
3. Place the sample in the bowl. Obtain the mass.
4. Cover the sample with trichlorethane. Allow it to soak (up to one hour).
5. Measure the mass of the filter ring. Place it on a bowl. Clamp a lid on the bowl.
6. Place the bowl in a centrifuge. Rotate it at a speed of up to 3600 rpm until the solvent ceases to flow.
7. Stop the centrifuge. Add about 200 ml of solvent. Rotate again.
8. Repeat the procedure (not less than three washings) until the extract is no longer cloudy and fairly light in color.
9. Place the bowl in the oven. Dry.
10. Brush the loose particles from the filter into the bowl.
11. Obtain the mass of the filter and bowl with dry aggregates.

Results:

Mass of bowl + sample	_____	g (A)
Mass of bowl	_____	g (B)
Mass of filter	_____	g (C)
After Test:		
Mass of bowl + sample	_____	g (D)
Mass of filter	_____	g (E)

Calculations:

Mass of sample (A – B)	_____	g (M)
Mass of aggregates in bowl (D – B)	_____	g (F)
Mass of aggregates in filter (E – C)	_____	g (G)
Mass of aggregates (F + G)	_____	g (M_G)

Mass of asphalt $(M - M_G)$ _____ g (M_B)
Asphalt content M_B/M _____ %

Note: A sieve analysis can be conducted on aggregates from the bowl to obtain the grain-size distribution curve for the aggregates in the mixture. The mass of aggregates in the filter (G) should be added to the amount passing the 75-μm (No. 200) sieve.

6-9 PROBLEMS

Note: Problems 6–9, 6–10, and 6–12 are repeated in traditional units as problems 6–41, 6–42, and 6–43.

6–1. The viscosity of an asphalt is 750 centistokes. Estimate its viscosity in these units: (a) cm^2/s; (b) seconds, Saybolt-Fural; (c) Pa $\cdot$ s; (d) poises.

6–2. A paving agency has been using 85/100 asphalt. What grade should it use in viscosity-graded asphalt?

6–3. What is liquid asphalt? How does it set or cure? What governs its rate of curing?

6–4. Recommend plant mixing temperature limits for asphalt C in fig. 6–2. If this asphalt is to be used as a seal coat by spraying on a pavement, what temperature would be required?

6–5. What is meant by MC 70? What are its main constituents? What is its viscosity at 60°C?

6–6. What are the constituents of asphalt cement? What causes deterioration or aging of asphalt in use? Why does this happen?

6–7. List all the viscosity requirements for AC 10 asphalt cement according to Table 1 of ASTM D 3381.

6–8. What is the thin film oven test? Why does it make sense to use this type of test?

6–9. An asphalt concrete mixture has a density of 2385 kg/m^3. The asphalt content is 6.8%, and asphalt absorption is 0.7%. Find the mass of the aggregate and the net asphalt in each cubic meter.

6–10. An asphalt concrete mix has a density of 2370 kg/m^3 at an asphalt content of 5.8%. The asphalt absorption is 0.9%. The relative density of the aggregates is 2.67; of the asphalt, 1.05. Find the percentages of air voids and VMA.

6–11. Water absorption of an aggregate is 1.3%. Estimate the asphalt absorption for this aggregate.

6–12. Asphalt content of a mix is 5.7%, expressed as a percentage of the mass of the aggregates. Find the percentage of air voids and voids in mineral aggregate if asphalt absorption is 0.5% and the relative densities of the aggregate and asphalt cement are 2.66 and 1.02 respectively. Density of the mix is 2390 kg/m^3.

6–13. What causes aging or cracking of asphalt concrete? How are aging and cracking controlled in the design, production, and construction of pavement?

6–14. What characteristics of an asphalt concrete mix are the most important in the development of strength?

6–15. Specifications for asphalt concrete often give a minimum and a maximum value for allowable air voids. Why? What property (or properties) is affected, and how?

6–16. Why is a minimum percentage for the VMA in an asphalt concrete mix usually specified? What property is affected and how?

6–17. An asphalt concrete specification requires that the asphalt plant be operated at a temperature corresponding to a viscosity of 225 centistokes ± 5°C. What would be the approved temperature range if the plant used the asphalt cement B described in fig. 6–2? Why are the upper and lower limits specified? In each case, what property is affected, and how?

6–18. What is mineral filler? Why is it important?

6–19. What properties of aggregates are important to ensure a high degree of skid resistance in a pavement surface? Why?

6–20. A sample of limestone dust (mass 104.3 g) is tested in a washed sieve analysis for use as mineral filler with the following results:

Sieve	Mass retained
600 μm (No. 30)	0
300 μm (No. 50)	0.66 g
150 μm (No. 100)	2.03 g
75 μm (No. 200)	13.17 g

Check for acceptance according to the AASHTO specification quoted.

6–21. The following aggregates are to be tested for acceptance for use in an asphalt concrete base course. Grading No. 1 is specified for the fine aggregate (table 6–5) and Size No. 57 (table 6–6) for the coarse. Report on acceptance according to specifications quoted.

(a) Mineral filler: Total sample tested = 114.3 g

Sample is washed through 600, 300 and 75 μm (No.

30, No. 50 and No. 200) sieves, then dried.

Mass retained on 600 μm (No. 30) =   0 g

300 μm (No. 50) =   8.8 g

75 μm (No. 200) = 24.5 g

Atterberg Limits Test results: w_L = 22

w_P = 20

(b) Fine aggregate: Original sample = 531.2 g

Washed sample = 504. 8 g

Dry sieving: retained on 9.5 mm (3/8-in)   0 g

4.75 mm (No. 4)   24.3 g

2.36 mm (No. 8)   105.3 g

1.18 mm (No. 16)   88.7 g

600 μm (No. 30)   117.9 g

300 μm (No. 50)   85.4 g

150 μm (No. 100)   43.7 g

75 μm (No. 200)   28.4 g

Pan   10.3 g

Soundness test ($mgSO_4$): original sample = 835.1 g

final sample   = 693.8 g

(c) Coarse aggregate

(crushed stone): Sample size 11321 g

Sieving test: retained on 38 mm (1½-in)   0 g

25 mm (1-in)   418 g

12.5 mm (1/2-in)   4604 g

9.5 mm (3/8-in)   2633 g

4.75 mm (No. 4)   3091 g

2.36 mm (No. 8)   416 g

Pan   148 g

Abrasion test: original sample = 5010 g

final sample   = 2608 g

Soundness test: original sample = 4377 g

final sample   = 4040 g

6–22. Name three properties of aggregates (other than gradation) that are important for asphalt concrete. Indicate why they are important.

6–23. The aggregates in problem 6–21 are to be used for a trial mix. The proportions are 55% coarse aggregate, 40% fine aggregate, and 5% mineral filler. The coarse aggregate is divided into three sizes for blending—over 12.5 mm (1/2 in), 12.5 mm-9.5 mm (1/2 in-3/8 in) and under 9.5 mm (3/8 in). The fine aggregate is divided into sizes larger than 1.18 mm (No. 16) and smaller than 1.18 mm (No. 16). How much from each of the three coarse fractions, the two fine fractions, and the mineral filler should be used for each briquette containing 1200 g of aggregates?

6–24. An asphalt briquette has a mass of 739.2 g in air and 411.7 submerged. Find the density in kg/m^3 and lb/ft^3.

6–25. Why is a higher value for maximum percentage of air voids allowed in base courses than in surface courses in the Asphalt Institute's suggested requirements for mixes evaluated by the Marshall method?

6–26. Following are gradation test results for four aggregates being considered for use in a base course asphalt concrete. Proportion these to produce a blend with a gradation approximately in the center of allowable range given in table 6–7. Also check if your combination of the two coarse aggregate sizes meets the requirements of the relevant ASTM specification for coarse aggregate.

Sieve	% Passing			
	Agg. A	Agg. B	Agg. C	Agg. D
38 mm (1½ in.)	100			
25 mm (1 in.)	93			
19 mm (3/4 in.)	53	100		
12.5 mm (½ in.)	8	91		
9.5 mm (3/8 in.)	3	65	100	
4.75 mm (No. 4)	1	12	96	
2.36 mm (No. 8)	0	5	75	
1.18 mm (No. 16)		2	50	
600 μm (No. 30)		0	38	
300 μm (No. 50)			26	100
150 μm (No. 100)			5	98
75 μm (No. 200)			2	88

6–27. Tests on a Marshall briquette yielded the results shown. The aggregate was made up of 45% aggregate A, 35% aggregate B, and 20% aggregate C with relative density (specific gravity) values of 2.72, 2.67 and 2.64, respectively. Calculate the density, stability, flow, percentage of air voids and VMA and compare with the Asphalt Institute's criteria for base courses on low volume highways. Nominal maximum size of the aggregates is 25 mm (1 in). The relative density (specific gravity) of the asphalt cement is 1.01, and the asphalt absorption is 1.0%.

Mixing data—mass of aggregate	1200 g
—mass of asphalt	58 g
Briquette data—mass in air	1241.3 g
—mass submerged	712.7 g
Stability test—maximum load	1870 N (420 lb.)
at a strain of	2.4 mm (0.094 in.)

6–28. Following are the results of a Marshall trial mix:

% AC	Density		Stability		Flow		Air voids	VMA
	kg/m^3	lb/ft^3	N	lb	mm	in	%	%
5	2365	147.6	2010	452	1.8	0.07	6.8	18.8
6	2388	149.0	2620	589	2.5	0.10	5.0	16.6
7	2398	149.6	2250	506	3.4	0.13	3.9	15.6
8	2386	148.9	1500	337	4.6	0.18	3.2	16.2

Plot the test results on graphs as shown in Fig. 6–12. Find the optimum asphalt content to meet suggested requirements for a surface course (12.5 mm or ½ in aggregate) for a medium traffic area. Record the properties of your design mix.

6–29. Following are the results of a Marshall trial mix for a surface course (12.5 mm or ½ in aggregate) being evaluated for use in a heavy traffic area:

% AC	Density		Stability		Flow		Air voids	VMA
	kg/m^3	lb/ft^3	N	lb	mm	in	%	%
5	2355	147.0	2210	497	2.8	0.11	7.3	19.6
6	2373	148.1	2500	562	3.7	0.15	6.1	18.0
7	2384	148.8	2380	535	4.6	0.18	5.0	17.2
8	2372	148.0	1900	427	5.9	0.23	4.6	18.3

Plot and analyze the test results. If it is not possible to find an optimum asphalt content for this trial, what changes in proportions of coarse and fine aggregate and mineral filler would you recommend for a second trial.

6–30. What are the main steps in production at an asphalt plant? How does the drum mixing method differ from traditional methods?

6–31. An asphalt batch plant produces a mixture using 6.4% asphalt cement, and aggregates A, B, and C in the proportions of 45:35:20. The batch size is 2400 kg (5000 lb). The gradation of the aggregates is:

	Aggregate A	Aggregate B	Aggregate C
Passing 25 mm (1-in)	100%		
19 mm (3/4-in)	88%		
12.5 mm (1/2-in)	62%		
9.5 mm (3/8)	32%	100%	
4.75 mm (No. 4)	8%	92%	100%

2.36 mm (No. 8)	0%	76%	80%
1.18 mm (No. 16)		51%	67%
600 μm (No. 30)		30%	54%
300 μm (No. 50)		19%	47%
150 μm (No. 100)		9%	37%
75 μm (No. 200)		2%	31%

Gradation of the material in the three hot bins is:

Coarse bin	Passing 25 mm (1 in)	100%
	9.5 mm (3/8 in)	5%
	2.36 mm (No. 8)	0%
Medium bin	Passing 9.5 mm (3/8 in)	100%
	2.36 mm (No. 8)	12%
Fine bin	Passing 2.36 mm (No. 8)	100%

Find the required mass of asphalt cement and aggregates from each bin per batch.

6–32. An asphalt extraction test records the following values:

Mass of bowl	3271.3 g
Mass of bowl + sample	4109.8 g
Mass of filter	10.5 g
After test: Mass of bowl and dry material	4045.3 g
Mass of filter	16.2 g

What is the asphalt content of this mixture? How much aggregate was trapped in the filter?

6–33. Discuss the drum-mixing concept for asphalt concrete. Why is it more economical than traditional plants? Why would the finished product be more durable than that from the traditional plants?

6–34. Specifications for a highway paving project require that the minimum temperature of the mix at the paver be that required for a viscosity of 10.0 cm^2/s (1000 centistokes). Why is this specified? What temperature would this be if asphalt C in fig. 6–2 is being used?

6–35. A contract requires that the asphalt concrete be compacted to a minimum density (unit weight) equal to 98% of its design density. What field density (unit weight) would be required for the mix described in section 6–4.5?

6–36. How much reclaimed material, as a percentage of the total mix, is the maximum amount usually included in recycled asphalt prepared in (a) drum plants, and (b) in traditional plants?

6–37. A highway department uses AC 5 (Pen. 120–150) for a certain type of highway. What grade should be added to recycled pavement material to produce a rejuvenated mixture? Why is this practice followed?

6–38. What is a surface treatment?

6–39. What grades of liquid asphalts would be suitable for surface treatments placed in cool weather with 12.5-mm (½-in) aggregate? Why can higher viscosity asphalts be used in warmer weather?

6–40. A surface treatment is proposed for a road 7.3 mi (11.7 km) in length and 24 ft (7.3 m) wide. ASTM grading 5 is to be used for the single-type treatment. Estimate, in tons (tonnes), the required quantities of asphalt and aggregates. The density of the aggregate as placed is 100 lb/ft^3 (1600 kg/m^3); the relative density of the asphalt is 1.0.

6–41. An asphalt concrete mixture has a unit weight of 148.8 lb/ft^3. The asphalt content is 6.8%, and asphalt absorption is 0.7%. Find the weight of aggregate and net asphalt in each cubic foot.

6–42. An asphalt concrete mix has a unit weight of 147.9 lb/ft^3 at an asphalt content of 5.8%. The asphalt absorption is 0.9%. The specific gravity of the aggregates is 2.67; of the asphalt, 1.05. Find the percentages of air voids and VMA.

6–43. Asphalt content of a mix is 5.7%, expressed as a percentage of the weight of the aggregates. Find the percentage of air voids and voids in the mineral aggregate if the asphalt absorption is 0.5% and the specific gravities of the aggregates and the asphalt cement are 2.66 and 1.02, respectively. Unit weight of the mix is 149.1 lb/ft^3.

7

Portland
Cement Concrete

Portland cement concrete is a concrete or artificial rock composed of aggregates, water, and a cementing agent known as portland cement. Portland cement is made from limestone (or some other source of lime) and other minerals, which are ground up, mixed, burned in a kiln, and subsequently ground to a fine powder which will harden when mixed with water.

7–1 PORTLAND CEMENT

The most important and most costly material in this type of concrete is the cementing agent, portland cement. It is widely manufactured, usually from readily available bedrock or shell deposits. The name *portland* is not a trade name but the name of a type of cementing material, in a similar way that asphalt, epoxy, lime, etc., are cementing materials. It was called portland cement because it resembled cement obtained by grinding natural rocks found on the Isle' of Portland.

7–1.1. The main minerals required for the production of portland cement are lime (CaO), silica (SiO_2), alumina (Al_2O_3), and iron oxide (Fe_2O_3). Table 7–1 lists some sources of these components. The main component is lime (60–65%), and therefore manufacturing plants are usually located to take advantage of one of the various sources of lime.

274

Table 7–1

TYPICAL SOURCES OF RAW MATERIALS USED IN MANUFACTURING OF PORTLAND CEMENT*

Lime	Silica	Alumina	Iron
Cement rock	Sand	Clay	Iron ore
Limestone	Traprock	Shale	Iron calcine
Marl	Calcium silicate	Slag	Iron dust
Alkali waste	Quartzite	Fly ash	Iron pyrite
Oyster shell	Fuller's earth	Copper slag	Iron sinters
Coquina shell		Aluminum ore refuse	Iron oxide
Chalk		Staurolite	Blast-furnace
Marble		Diaspore clay	flue dust
		Granodiorite	
		Kaolin	

*Courtesy of the Canadian Portland Cement Association.

7–1.2 In the manufacture of portland cement, the raw materials are ground up, mixed to produce the desired proportion of minerals, and burned in a large kiln. The temperature in the kiln reaches about $1500°C$ ($2700°F$). This drives off water and gases and produces new chemical compositions in particles called *clinker*. The clinker is subsequently ground with about 5% gypsum (to control rate of hardening) and is ready for use as portland cement.

Figure 7–1 illustrates the manufacturing process for portland cement in a typical plant.

The cement compounds produced are:

— $3CaO \cdot SiO_2$, abbreviated C_3S
— $2CaO \cdot SiO_2$, abbreviated C_2S
— $3CaO \cdot Al_2O_3$, abbreviated C_3A
— $4CaO \cdot Al_2O_3 \cdot Fe_2O_3$, abbreviated C_4AF.

The relative amounts of these four chemicals in the final product depend on the desired properties such as rate of hardening, amount of heat given off, and resistance to chemical attack.

Portland cement is a hydraulic cement; that is, one that sets or hardens when mixed with water. Particles of cement take up water, forming a gel which cements the individual particles together. This chemical process is called *hydration*. It continues for months or years as long as water is present, as shown in fig. 7–2. The solid line shows that concrete, which reaches 100% of its design strength in 28 days, continues to increase in strength, reaching about 120% and 130% of design strength in 90 days and 180 days, respectively.

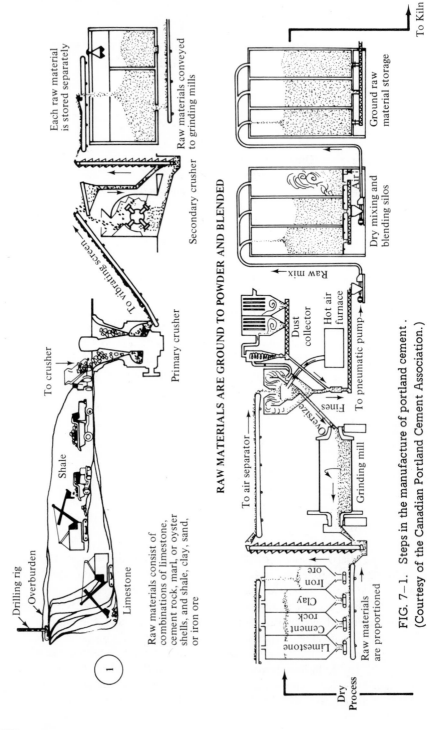

STONE IS FIRST REDUCED TO 5-IN. SIZE, THEN 3/4-IN., AND STORED

RAW MATERIALS ARE GROUND TO POWDER AND BLENDED

FIG. 7-1. Steps in the manufacture of portland cement.
(Courtesy of the Canadian Portland Cement Association.)

RAW MATERIALS ARE GROUND, MIXED WITH WATER TO FORM SLURRY, AND BLENDED

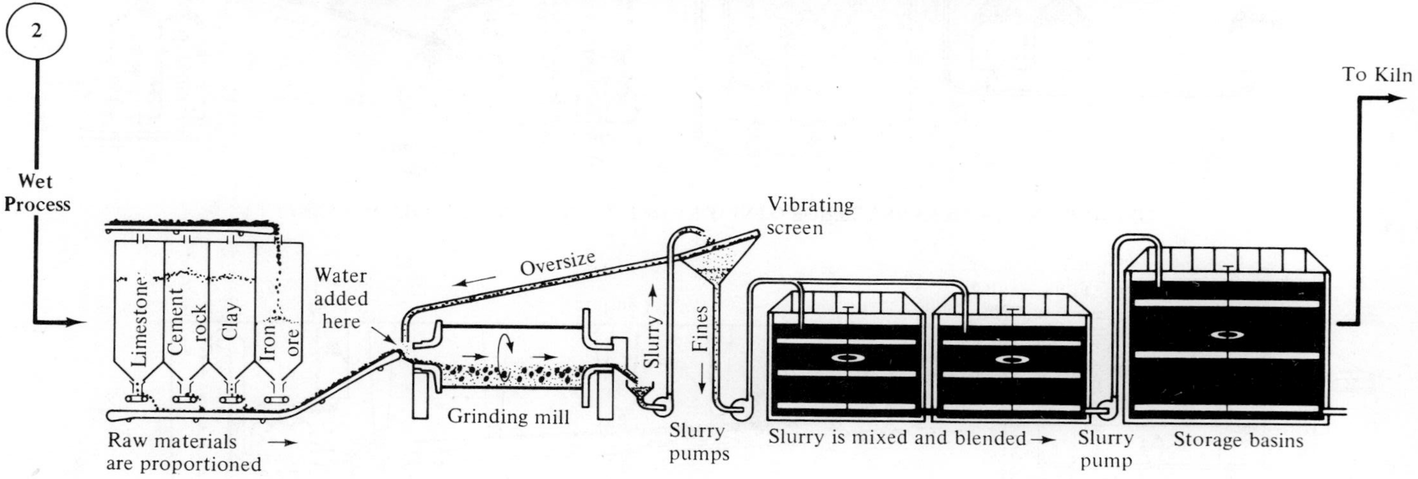

FIG. 7–1. continued

BURNING CHANGES RAW MIX CHEMICALLY INTO CEMENT CLINKER

CLINKER WITH GYPSUM ADDED IS GROUND INTO PORTLAND CEMENT AND SHIPPED

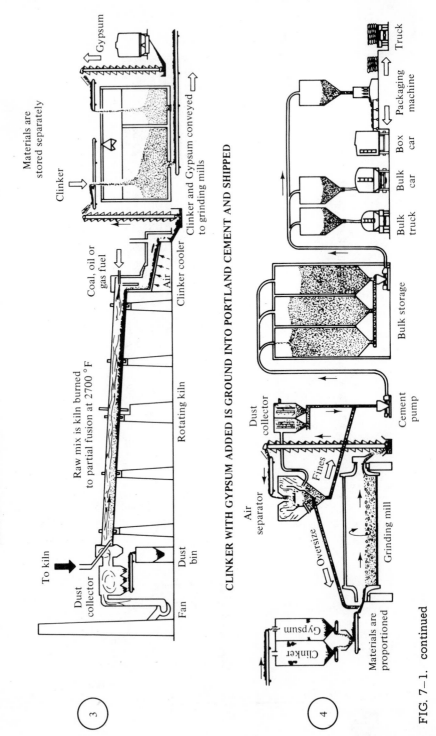

FIG. 7–1. continued

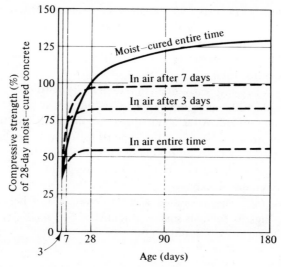

FIG. 7-2. Variation of concrete strength with age and curing conditions. (Courtesy of the Canadian Portland Cement Association.)

The chemical reaction for C_3S is

$$2C_3S + 6H_2O \rightarrow 3CaO \cdot 2SiO_2 \cdot 3H_2O + 3Ca(OH)_2$$

The main product, $3CaO \cdot 2SiO_2 \cdot 3H_2O$, is *calcium silicate hydrate* or *tobermorite*, and gives concrete its strength. The chemical reactions for the other cement compounds are similar.

The total amount of water required to complete the hydration of the cement is about 25% of the mass of the cement.

C_3S hardens rapidly and is mainly responsible for the initial set and early strength. C_2S hydrates slowly and is the main source of increased strength after the first week of hardening. These two compounds are mainly responsible for the strength of the concrete. C_3A reacts very quickly and adds a small amount of strength. C_4AF reacts slowly. Its main purpose in portland cement is to reduce the temperature required during burning in the kiln.

During hydration, heat (called *heat of hydration*) is given off. In massive structures, such as dams, this heat could result in cracking due to the fairly rapid rise in temperature and the subsequent cooling. The rate of release of heat is greatest in C_3A, followed by C_3S. About 50% of the total heat of hydration is released in the first three days.

Sulphates are often found in soils and groundwater. These may combine with C_3A to produce an expanding compound which could result in disintegration of the concrete.

Characteristics of these four compounds are summarized in table 7–2.

Five main types of portland cement are usually produced. The typical compositions and the usual designations are shown in table 7–3.

Normal portland cement (Type I or 10) is the most common, accounting for 90%–95% of the total.

In cases where high early strength is desired, Type III (30) may be used. It usually costs 10%–20% more than Type I (10), but it is about 90% stronger one day after being poured. After about three months, both concretes would be about equal in strength. Note that the Type III (30) cement contains more C_3S, the component that hardens most rapidly and provides early strength. This type of cement is also ground finer to allow the water to reach the interior of the cement particles more quickly.

Type IV (40) cement is used where a lower rate of heat increase is required, as in dams. It contains smaller amounts of C_3S and C_3A, both of which harden rapidly and give off heat quickly.

Table 7–2
CHARACTERISTICS OF THE PORTLAND CEMENT COMPOUNDS

	C_3S	C_2S	C_3A	C_4AF
Rate of hydration	Medium	Slow	Fast	Slow
Strength—early	High	Low	Medium	Low
—ultimate	High	High	Low	Low
Amount of heat liberated	Medium	Low	High	Low
Resistance to chemical attack	Good	Good	Poor	Good

Table 7–3
TYPICAL COMPOUND COMPOSITION OF PORTLAND CEMENT*

Designations of Portland Cement			Compound Composition (%)			
USA	Canada	Type	C_3S	C_2S	C_3A	C_4AF
I	10	Normal	50	24	11	8
II	20	Moderate	42	33	5	13
III	30	High Early Strength	60	13	9	8
IV	40	Low Heat of Hydration	26	50	5	12
V	50	Sulfate-Resisting	40	40	4	9

*Courtesy of the Canadian Portland Cement Association.

Type V (50) cement should be used in cases where the groundwater contains over 1000 ppm of sulphate, or where the soil contains over 0.2% sulphate. The amount of C_3A in this type is only about one-third of that in normal cement.

Type II (20) cement is suitable where moderate resistance to sulphate (over 150 ppm in the water or 0.1% in the soil) or slightly lower heat of hydration is required.

The relative strengths of these cements are indicated in table 7−4.

Types IA, IIA, and IIIA are similar to Types I, II, and III but contain small amounts of air entraining materials to improve the durability of the concrete.

Slag cements and pozzolan cements contain ground slag or pozzolanic materials with the portland cement for economy or for special uses. Concrete made with these cements is described in section 7−10.

Other special cements, for example, white portland cement and masonry cement, are also produced.

Table 7−4

APPROXIMATE RELATIVE STRENGTHS OF CONCRETE

Type of Portland Cement	Compressive Strength (% of Normal Portland Cement Concrete)				
	1 Day	3 Days	7 Days	28 Days	3 Months
I. 10	100	100	100	100	100
II. 20	75	80	85	90	100
III. 30	190	190	120	110	100
IV. 40	55	55	55	75	100
V. 50	65	65	75	85	100

7−1.3 Some of the more important properties and tests used to check the quality of portland cement are:

1. *Fineness*—This helps govern the rate of hydration because smaller particles will absorb water faster and hydrate sooner.
2. *Setting*—Tests are conducted to measure the setting time.
3. *Compressive strength*—Fifty-millimeter (2-in) concrete cubes are made with standard sand to measure compressive strength.
4. *Tensile strength*—Standard molds are used to produce samples for measuring tensile strength. This is usually about 10% of the compressive strength of the cement.
5. *Relative density (specific gravity)*—This is usually 3.15 for portland cement.

7–2   *PROPERTIES OF PORTLAND CEMENT CONCRETE*

Portland cement concrete, hereafter called *concrete*, is a mixture of portland cement, aggregates, and water. Strength is the main requirement, and concrete design is usually based on this property. However, durability, workability, and other properties are also very important. For pavements, in particular, durability requirements may govern the quality of concrete.

7–2.1   Portland cement concrete is made up of:

$$\left.\begin{array}{l}\text{Portland cement} \\ \text{Water} \\ \text{Air}\end{array}\right\} \text{Paste}$$

$$\left.\begin{array}{l}\text{Fine aggregate} \\ \text{Coarse aggregate}\end{array}\right\} \text{Aggregate}$$

Typical proportions are shown in fig. 7–3.

Two factors that are extremely important for the quality of the concrete are (1) the water/cement ratio (W/C) and (2) whether or not the concrete is air entrained.

7–2.2   The water/cement ratio is the mass of the water in the mix divided by the mass of the cement, expressed as a ratio. For example, a m³ batch may con-

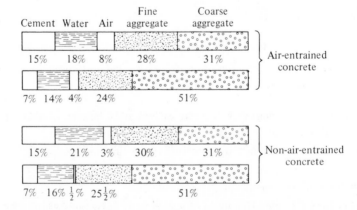

FIG. 7–3. Range in proportions of materials used in concrete. (Courtesy of the Canadian Portland Cement Association.)

tain 190 kg of water and 340 kg of cement. The water/cement ratio (W/C) is 190/340 = 0.56. For most mixtures in everyday use, the W/C ratio is between 0.4 and 0.7.

The water-cement relationship can also be measured in units of gallons of water per sack of cement. A W/C ratio of 0.50 corresponds to 5.64 U.S. gallons (8.33 lb) per sack (94 lb).

Water is required in the mixture for two purposes: (1) to react chemically with the cement and cause it to harden (this is the water required for hydration); and (2) to make the mix plastic or workable enough to be used as intended. Because about 1 kg of water is required for 4 kg of cement for hydration, a W/C ratio of 0.25 would be sufficient to harden the mixture. However, with this amount of water the mix would not be plastic. It would act as a dry mass of aggregates, and could not be molded, moved, or finished properly. Therefore, extra water is required to make the concrete plastic.

This extra water, although required for workability, is detrimental to the properties of the hardened concrete. Consider mixes A and B with typical proportions of the paste fraction for 1 m^3 of concrete:

	A	B
Cement (kg)	400	400
Water (kg)	240	160
W/C ratio	0.6	0.4
Water required for hydration (kg)	100	100
Excess water (kg)	140	60

Both mixes require 100 kg of water for hydration because they both contain 400 kg of cement. Mix A, with substantially more water, will flow more easily and be more workable. However, as the concrete sets, the excess water will form water voids in the mix. As the concrete dries and hardens, this water may evaporate, leaving air voids in the concrete. These water or air voids will weaken concrete substantially. Mix B has less than half of the excess water that is in mix A and therefore will be denser and much stronger.

The strength of concrete is determined to a large degree by the strength of the paste. The volume of the excess water in Mix A is about 38% of the paste volume, and only about 21% in Mix B. (Volume calculations are described in section 7−6.4.)

7−2.3 One of the most important advances in concrete technology in recent years has been the use of entrained air to protect against disintegration resulting from the freezing and thawing of water in pores in the concrete.

Weathering or disintegration due to freezing and thawing is caused by the expansion of the water as it freezes. The pressure caused by this expansion forces the pore open. After thawing, the larger pore is resaturated with water and subsequent freezing increases the pore volume again. This pressure can soon build up to exceed the tensile strength of the concrete and cause pieces of concrete to break off at the surface. This is the most common cause of disintegration of exposed concrete.

In air-entrained concrete, small bubbles of air, approximately 0.025 mm–0.075 mm (0.001 in–0.003 in) in diameter, are formed in the concrete by adding special materials, called *air-entaining agents,* to the mix. A very large number of these bubbles, which are not connected, is formed. The cushioning effect of the air bubbles relieves the pressure developed by the freezing of water in pores. Some water flows into these bubbles, and, on thawing, capillary tension causes this water to flow back to its original location.

All concrete contains some air. However, in non-air-entrained concrete, this entrapped air is made up of relatively large bubbles that are not effective in resisting the weathering process.

Figure 7–4 illustrates the resistance of air-entrained concrete to weathering. All concrete that will be exposed to freezing and thawing conditions should be air entrained. To be effective, the entrained air should be distributed uniformly throughout the volume between the coarse aggregate particles. The air bubbles should be spaced closely so that the pressure can be relieved before the tensile strength of the concrete is reached. This spacing may be about 0.2 mm (0.008 in).

Experience has shown that about 9% air in the mortar fraction (paste plus fine aggregate) provides adequate protection, except for concrete subjected to deicing chemicals, in which case a higher percentage is required. Concrete made with small-size coarse aggregates requires more mortar to fill the spaces between the coarse particles. Therefore, air as a proportion of the whole mix increases as the size of the largest particles in the mix decreases. Table 7–5 indicates how, theoretically, the air content varies with aggregate size.

The billions of disconnected air bubbles in air-entrained concrete also increase workability of the mix and reduce both the segregation of the aggregates from the paste and the bleeding of the paste to the surface. The air bubbles act as minute ball bearings, aiding flow of the material as a unit.

7–2.4 Some of the more important properties of hardened concrete are compressive strength, tensile strength, durability, permeability, and hardness or abrasion resistance.

Compressive strength of concrete, expressed in MPa (psi), is usually measured by the load required to break a cylinder that has been cured for 28 days; it is obtained by dividing the total failure load by the cross-sectional area. The use of 28 days as the age at which concrete is assumed to have reached its design

(a) Non-air-entrained

(b) Air-entrained

FIG. 7–4. Effect of severe weathering on non-air-entrained and air-entrained concrete. (Courtesy of the Portland Cement Association.)

Table 7—5

VARIATION IN AIR REQUIREMENT WITH SIZE OF AGGREGATE

Maximum Size of Coarse Aggregates (mm) (in.)		Approximate Percentage of Mortar in Mix	Total Air Content (% of Mortar)
75	3	50	4.5
38	1½	55	5.0
19	¾	60	5.4
9.5	3/8	75	6.8

strength is standard. For normal cement concrete, the strength at 3, 7, and 14 days is about 40, 60, and 75%, respectively, of the 28-day strength.

As shown earlier in fig. 7—2, concrete will continue to increase in strength as long as water is present, although at a decreasing rate. Note that strength gain ceases as soon as the concrete dries out after being exposed to air because water is required for hydration to take place.

Compressive strength will also vary considerably with the W/C ratio, as noted earlier. Figure 7—5 illustrates this variation for air-entrained and non-air-entrained concrete. From these graphs it would appear that the use of air-entrained concrete results in a substantial reduction in strength. For example, at W/C = 0.50, the 28-day strengths are about 33 MPa for non-air-entrained concrete and 25 MPa for an air-entrained mix. However, air entrainment reduces the amount of water required for a given workability, thus allowing a reduction of the W/C ratio. To illustrate, a non-air-entrained concrete might require 200 kg of water to 400 kg of cement to be workable (W/C = 0.50). However, with air-entrained concrete the same degree of workability could be obtained with a mix containing about 175 kg of water and 400 kg of cement, with a resulting water/cement ratio of 0.44. The estimated strength of this mix would be about 29 MPa, not too far below the 33 MPa expected for non-air-entrained concrete containing the same amount of cement.

Tensile strength of concrete is very low—less than 10% of the compressive strength. It varies with compressive strength. The tensile strength is not considered in structural concrete, for reinforcing steel is used to carry tensile stresses. However, cracking of concrete due to volume change is usually due to tensile failure.

Flexural strength, or *modulus of rupture,* is often used to evaluate the strength of pavement concrete. A small beam is cast and loaded on one side until failure occurs due to tensile stress in the bottom of the beam. Flexural strength is usually about 15% of the compressive strength for ordinary concrete. A more accurate relationship is:

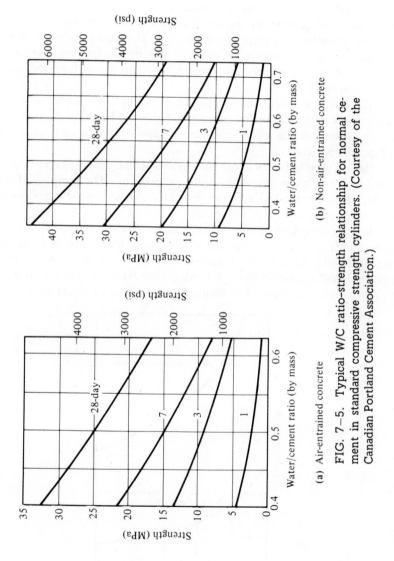

(a) Air-entrained concrete

(b) Non-air-entrained concrete

FIG. 7–5. Typical W/C ratio–strength relationship for normal cement in standard compressive strength cylinders. (Courtesy of the Canadian Portland Cement Association.)

$$\text{Flexural strength} = K \sqrt{\text{compressive strength}}$$

where K is a constant, approximately 0.7 for strengths in MPa, and 8.4 with psi values.

Durability of concrete is its resistance to various conditions causing disintegration. Certain types of aggregates termed *reactive aggregates* may cause problems. These are discussed in section 7—3. The main conditions affecting durability of exposed concrete are cycles of freezing and thawing temperatures. The effect of air entrainment and the W/C ratio on freeze—thaw resistance is illustrated in fig. 7—6.

The effect of air-entrainment in improving the resistance of concrete to disintegration due to cycles of freezing and thawing has been discussed. The strength of the concrete is also important, and low W/C ratios are recommended for exposed concrete.

Deicing chemicals such as calcium chloride (commonly used on pavements, sidewalks, etc.) frequently cause scaling or break-up. The presence of these chemicals increases the hydraulic pressure in the concrete, and stronger concrete with a higher air content is recommended.

In some areas groundwater or soil contains a high proportion of sulphates. These sulphates attack concrete and may cause disintegration. Sea water can bring about the same result. Concrete exposed to sulphates should be air entrained as well as having a lower W/C ratio. Type II (20) or type V (50) cement may also be required. Figure 7—7 illustrates the effect of air entrainment and the

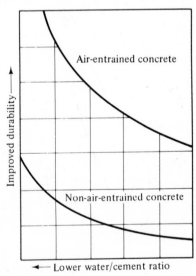

FIG. 7—6. Relationship between freeze—thaw resistance and W/C ratio for air-entrained and non-air-entrained concrete. (Courtesy of the Canadian Portland Cement Association.)

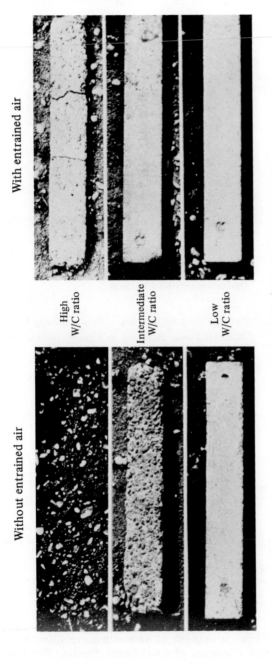

With entrained air

Without entrained air

High W/C ratio

Intermediate W/C ratio

Low W/C ratio

FIG. 7-7. Effect of entrained air and the W/C ratio on the durability of concrete specimens exposed to sulphate soil. (Courtesy of the Portland Cement Association.)

W/C ratio on the performance of concrete specimens subjected to sulphate attack.

Permeability or watertightness also varies greatly with the W/C ratio. Concrete with a high W/C ratio will have more water and/or air voids, and therefore will be more permeable. Results of standard leakage tests using 25-mm-thick discs of concrete are shown in table 7–6.

Table 7–6

LEAKAGE THROUGH CONCRETE SPECIMENS*

W/C Ratio	Leakage (kg/m^2 per h)	
	7 Days	28 Days
0.50	0	0
0.64	5	0
0.80	12	1

*Courtesy of the Canadian Portland Cement Association (modified by the author).

Abrasion resistance varies greatly with concrete strength, as would be expected. Figure 7–8 indicates wear resulting in a standard test on concrete of various strengths.

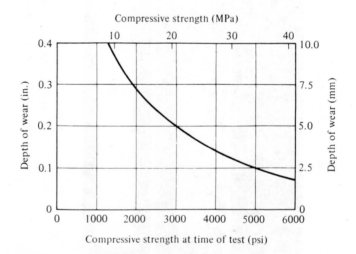

FIG. 7–8. Variation of depth of wear with compressive strength in an abrasion test. (Courtesy of the Canadian Portland Cement Association.)

7–2.5 Important properties of plastic concrete include workability and harshness.

Workability indicates the consistency or plasticity or ease of placing and molding the concrete without segregation. Generally, increased water makes concrete more workable. Air entrainment also increases workability.

Concrete should be workable, but it should not segregate or bleed. Bleeding is the movement of water to the surface. This may increase the W/C ratio near the surface which can result in a weak layer, particularly if the concrete is finished with this excess water on the surface. Bleeding is caused by the settling of the solids in the fresh concrete. Segregation is the separation of coarse aggregates from the mortar. Both of these problems may be caused by excessive vibration during placement, too much water in the mix, or poor mix design. The concrete mix should be designed with the minimum water required for placing, and it should be deposited as close as possible to the final location of the material in the structure so that it does not have to be moved by vibrators.

Workability is usually measured by the slump test.

Harshness is a property that indicates the finishing quality of the concrete. A harsh mix will have too much coarse aggregate or too coarse a sand, and it will be hard or impossible to finish smoothly. On the other hand, an over-sanded mix will appear fatty, and while it will finish well, it will be uneconomical.

7–2.6 Another important property of concrete that must be considered in design is volume change. Three types of volume change are of importance in this discussion.

Temperature causes changes in volume of concrete, as it does for steel and most other materials. The amount of change varies with the type of aggregate used, but an average value for the coefficient of expansion is 10 μm/m (or 10 millionths, 10×10^{-6} units per unit) per °C (5.5 μm/m or 5.5 millionths per °F).

Example 7–1: If the temperature increases by 35°C from winter to summer, find the change in length of a 20-m slab.

Increase in length (temperature is increasing)
20 m $\times$ 10 μm/m $\times$ 35 = 7000 μm = 7 mm

Another type of volume change is the shrinkage that occurs as the concrete dries out during curing (unless it is kept continually under water). The amount of this shrinkage varies approximately with the water content of the mixture. For plain concrete the values range from 400 to 800 μm/m (millionths) (the higher value applying to mixes with a high water content) when exposed to air at a relative humidity of 50%.

Example 7–2:   Using the average value (600 μm/m), find the drying shrinkage for a 30-m concrete slab.

$$\text{Shrinkage} = 30 \text{ m} \times 600 \ \mu\text{m/m} = 18000 \ \mu\text{m} = 18 \text{ mm}$$

About one-third of this shrinkage occurs in the first month and about 90% in the first year.

Concrete that is exposed to cycles of wetting and drying will expand and contract with the change in moisture conditions.

In reinforced concrete the rate of drying shrinkage is reduced to about 200 to 300 μm/m (millionths).

The third major type of volume change in concrete is creep. This is a change in volume caused by a continuously applied load. Figure 7–9 indicates the magnitude of volume change that might be expected. While volume change due to creep is especially important in pre-stressed concrete, it is not significant for pavement slabs.

7–3   MATERIALS

Materials used in portland cement concrete are portland cement, water, fine aggregate, coarse aggregate, and admixtures. Each of these must meet certain requirements in order that quality concrete be produced.

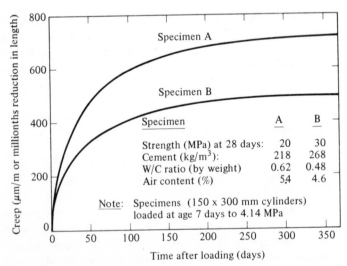

FIG. 7–9.   Volume change due to creep in test specimens. (Courtesy of the Canadian Portland Cement Association.)

7–3.1 Portland cement is manufactured in very large plants under conditions of close quality control. Specifications are given in ASTM Standard C150 and CSA Standard A5. Table 7–7 lists the main physical requirements for these cements.

7–3.2 Any municipal water or any natural water that is drinkable and has no definite taste or odor is satisfactory as mixing water. Seawater containing up to 35,000 ppm of dissolved salts is also suitable for unreinforced concrete. Early strength would be higher with salt water, but the 28-day strength might be lower. Using salt water in reinforced concrete could lead to corrosion of the reinforcing bars. A lower W/C ratio (maximum 0.45) and a thicker cover over the bars are often recommended if the concrete is exposed to seawater.

Water with other chemicals, organic material, industrial wastes, dirt, oil, or other impurities may be acceptable if concentrations are not too great. Tests should be made, and if the strength of test cubes made with the water is 90% of that obtained with pure water, the water is generally satisfactory.

7–3.3 Aggregates should be composed of clean, hard, strong, and durable particles. Important characteristics (other than gradation) include the following:

— hardness or resistance to wear
— soundness or resistance to freezing and thawing
— chemical stability
— particle shape and texture
— relative density and absorption
— deleterious substances

Aggregates must be hard, especially those used for pavement surfaces where skid resistance is very important, because soft aggregates tend to shine with traffic.

Soundness or resistance to freezing and thawing is necessary for concrete durability.

Chemical stability is important in concrete because certain aggregates, called *reactive aggregates,* react with alkalies in portland cement to form a gel around the particles. This gel may cause abnormal expansion and map-cracking of the concrete. Various special tests are used to identify possible reactive aggregates. Judging the performance of an aggregate in previous concrete work is a common method used to ensure selection of nonreactive aggregates. Tests involving the measurement of expansion of a mortar bar are sometimes conducted on unknown aggregates. Where reactive aggregates are found, special low alkali cements may be used or pozzolans may be added to the concrete. Most

Table 7-7

PHYSICAL REQUIREMENTS FOR PORTLAND CEMENTS
(Modified from ASTM C150)*

Cement Type	I	IA[a]	II	IIA[a]	III	IIIA[a]	IV	V
Air content of mortar, volume percent:								
Maximum	12	22	12	22	12	22	12	12
Minimum	–	16	–	16	–	16	–	–
Fineness, specific surface, cm²/g (alternative methods):								
Turbidimeter test, minimum	1600	1600	1600	1600	–	–	1600	1600
Air permeability test, minimum	2800	2800	2800	2800	–	–	2800	2800
Autoclave expansion, maximum percentage	0.80	0.80	0.80	0.80	0.80	0.80	0.80	0.80
Strength, not less than the values shown for the ages indicated below:								
Compressive strength, psi (MPa)								
1 day	1800 (12.4)	–	–	–	1800 (12.4)	1450 (10.0)	–	–
3 days	2800 (19.3)	1450 (10.0)	1500 (10.3)	1200 (8.3)	3500 (24.1)	2800 (19.3)	–	1200 (8.3)
7 days	–	2250 (15.5)	2500 (17.2)	2000 (13.8)	–	–	1000 (6.9)	2200 (15.2)
28 days	–	–	–	–	–	–	2500 (17.2)	3000 (20.7)
Time of setting (alternative methods):								
Gillmore test								
Initial set, minutes; not less than	60	60	60	60	60	60	60	60
Final set, hours; not more than	10	10	10	10	10	10	10	10
Vicat test:								
Initial set, minutes; not less than	45	45	45	45	45	45	45	45
Final set, hours; not more than	8	8	8	8	8	8	8	8

*Reprinted by the permission of the American Society for Testing and Materials, 1916 Race Street, Philadelphia, PA 19103, Copyright.
[a] Air entrained.

portland cements contain about 1% alkalies, Na_2O and K_2O. This is reduced to 0.6% or less in low alkali cements.

The use of long, thin aggregate particles should be restricted because they require more water to produce a workable mix and they break easily.

Relative density and absorption and the amount of surface moisture are important in mix design and corrections. Mix designs are based on aggregates in the saturated, surface dry condition.

Deleterious substances include dust coatings (finer than 0.075 mm or No. 200 sieve), coal, chert, soft particles, clay, shale, friable particles, and organic coatings. Deleterious actions of these substances are indicated in table 7–8.

A property of the coarse aggregate often considered in mix design is the dry rodded density. This is obtained by placing the aggregate in layers in a container and rodding each layer a specified amount. The dry rodded density is the total mass divided by the volume of the container.

Specifications for physical requirements for aggregates are given in table 7–9 (CSA) and fig. 7–10 (ASTM).

7–3.4 Gradation of aggregates is important for workability, strength, and economy.

For fine aggregates a minimum amount passing the 300-μm (No. 50) and 150-μm (No. 100 sieves), usually 10% and 2% respectively, is required to ensure that the surface can be finished easily to an acceptable degree of smoothness. Maximum amounts of these fine-sand sizes are also specified because the cement

Table 7–8
DELETERIOUS SUBSTANCES IN AGGREGATES*

Deleterious Substances	*Action on Concrete*
Organic impurities	Affect setting and hardening, may cause deterioration
Materials finer than 75-μm sieve	Affect bond, increase water requirement
Coal, lignite, or other low-density materials	Affect durability, may cause stains and popouts
Soft particles	Affect durability
Clay lumps and friable particles	Affect workability and durability, may cause popouts

*Courtesy of the Canadian Portland Cement Association.

Table 7–9

AGGREGATES FOR CONCRETE–CSA

a) Fine aggregate.
  i) Gradation limits.

Sieve Size	Total Passing Sieve Percentage by Mass
10 mm	100
5 mm	95–100
2.5 mm	80–100
1.25 mm	50– 90
630 μm	25– 65
315 μm	10– 35
160 μm	2– 10

(Not over 45% between any two consecutive sieves)
  ii) Fineness modulus between 2.3 and 3.1.
  iii) Free from injurious amounts of organic material.
  iv) Maximum loss in the soundness test ($MgSO_4$)–16%
  v) Maximum amount of deleterious substances–clay lumps, 1.0%; coal or lignite, 0.5%; fines, 3.0% (may be increased to 5.0% if not over 1% finer than 2 μm).
b) Coarse aggregate.
  i) Gradation limits (see p. 297).
  ii) Maximum loss in the abrasion test–50% (35% for floors and pavements).
  iii) Maximum amount of fines (smaller than 80 μm)–1.0% (may be increased to 1.5%, for crushed rock if free of shale and clay).
  iv) Maximum loss in soundness test ($MgSO_4$)–exposures A and B*–12%
  –exposures C and D*–18%
  v) Maximum amounts of deleterious substances–

	Clay lumps	Lightweight particles
Exposures A and B*	0.25%	0.5%
Exposures C and D*	0.5%	1.0%

c) Other requirements regarding particle quality and aggregate reactivity.

* See table 7–11 for exposure conditions.

(This information is adapted from CSA Standard CAN 3-A23.1–M 77, Concrete Materials and Methods of Concrete Construction, which is copyrighted by the Canadian Standards Association. Copies may be purchased from the association at 178 Rexdale Boulevard, Rexdale, Ontario, M9W 1R3.)

Nominal Size of Aggregate mm	Total Passing Each Sieve, Percent by Mass								
	56 mm	40 mm	28 mm	20 mm	14 mm	10 mm	5 mm	2.5 mm	1.25 mm
40-5	100	95–100	–	35–70	–	10–30	0–5	–	–
28-5	–	100	95–100	–	30–65	–	0–10	0–5	–
20-5	–	–	100	90–100	–	25–60	0–10	0–5	–
14-5	–	–	–	100	90–100	45–75	0–15	0–5	–
10-2.5	–	–	–	–	100	85–100	10–30	0–10	0–5

paste has to be sufficient to cover each particle, and an excessive quantity of fines would result in an uneconomical mix.

The fineness modulus of fine aggregate is the sum of the cumulative percentages retained on the 9.5-mm, 4.75-mm, 2.36-mm, 1.18-mm, 6.00-μm, 300-μm, and 150-μm sieves (3/8-in and Nos. 4, 8, 16, 30, 50, and 100, respectively) divided by 100. A sample calculation is as follows:

Pass Sieve	Cumulative % Passing	Cumulative % Retained
9.5 mm (3/8 in.)	100	0
4.75 mm (No. 4)	98	2
2.36 mm (No. 8)	91	9
1.18 mm (No. 16)	63	37
600 μm (No. 30)	38	62
300 μm (No. 50)	15	85
150 μm (No. 100)	4	96

Sum 291

The fineness modulus is 291/100 = 2.91. Note that the 75-μm (No. 200) size is not included in the calculation of the fineness modulus.

The fineness modulus is usually specified in addition to the gradation limits to ensure that an aggregate does not fall close to the coarse or the fine side of the limits for most or all of the sieve sizes. If an aggregate were close to the coarse limits on all sieves, even though it might meet the specification requirements, it might produce a very harsh or unworkable mixture. The fineness modulus is also useful in choosing proportions in trial mixes and in monitoring the consistency of production of sand from a plant over a period of time.

Various coarse aggregate gradations are allowed, depending on the maximum size of particles that can be used in the structure. Generally, the larger the allowable size, the more economical the concrete is because there will be less void space between aggregate particles to be filled with the expensive paste containing portland cement. However, maximum aggregate size must be limited to ensure that the concrete can be placed in the form without leaving voids or honeycombed areas.

Figure 7–11 indicates the effect of aggregate size on water requirements and therefore on cement requirements.

The importance of these and other properties of aggregates was discussed in some detail in Chapter 4. Test procedures and calculations required to measure these properties were also covered.

Figure 7–10 (ASTM) and Table 7–9 (CSA) give the main specifications used in the United States and Canada, respectively, for gradation and quality of concrete aggregates. Most highway departments have similar requirements.

Note that in the ASTM coarse aggregate gradation requirements, sizes 357, 467, 57, 67, 7 and 8, with nominal sizes extending to 4.75 mm (No. 4 sieve), are the usual gradations specified. The other gradations are used for special concrete or are blended with one of sizes noted.

7–3.5 Admixtures are used in concrete to improve its properties, to aid in construction procedures, to provide economy, and to fulfill other special pur-

1. *Fine aggregate:*

(a) Gradation limits.

Sieve	Percentage Passing*
3/8 in. (9.5-mm)	100
No. 4 (4.75-mm)	95 to 100
No. 8 (2.36-mm)	80 to 100
No. 16 (1.18-mm)	50 to 85
No. 30 (600-μm)	25 to 60
No. 50 (300-μm)	10 to 30
No. 100 (150-μm)	2 to 10

*Not over 45% retained between any two consecutive sieves

(b) Fineness modulus between 2.3 and 3.1.

(c) Free from injurious amounts of organic material

(d) Maximum loss in soundness test — 10% with sodium sulphate
15% with magnesium sulphate.

(e) For exposed concrete, aggregates shall not be reactive.

(f) Deleterious materials.

Item	Weight percent of total sample (max.)
Clay lumps and friable particles	3.0
Material finer than No. 200 (75-μm) sieve:	
Concrete subject to abrasion	3.0**
All other concrete	5.0**
Coal and lignite:	
Where surface appearance of concrete is of importance	0.5
All other concrete	1.0

** In the case of manufactured sand, if the material finer than the No. 200 (75-μm) sieve consists of the dust of fracture, essentially free from clay or shale, these limits may be increased to 5 and 7%, respectively.

FIG. 7–10. Specification for concrete aggregate. (Modified from ASTM Standard C33 by permission of the American Society for Testing and Materials, 1916 Race Street, Philadelphia, PA 19103, Copyright.)

2. *Coarse aggregate:*

(a) Grading limits.

Size Number	Nominal Size (Sieves with Square Openings)	Amounts Finer than Each Laboratory Sieve (Square Openings), Weight Percent												
		4 in. (100 mm)	3½ in. (90 mm)	3 in. (75 mm)	2½ in. (63 mm)	2 in. (50 mm)	1½ in. (38.1 mm)	1 in. (25.0 mm)	¾ in. (19.0 mm)	½ in. (12.5 mm)	⅜ in. (9.5 mm)	No. 4 (4.75 mm)	No. 8 (2.36 mm)	No. 16 (1.18 mm)
1	3½ to 1½ in. (90 to 37.5 mm)	100	90 to 100	...	25 to 60	...	0 to 15	...	0 to 5	...	...	...	...	...
2	2½ to 1½ in. (63 to 37.5 mm)	...	...	100	90 to 100	35 to 70	0 to 15	...	0 to 5	...	...	...	...	...
357	2 in. to No. 4 (50 to 4.75 mm)	...	...	...	100	95 to 100	...	35 to 70	...	10 to 30	...	0 to 5	...	...
467	1½ in. to No. 4 (37.5 to 4.75 mm)	...	...	...	...	100	95 to 100	...	35 to 70	...	10 to 30	0 to 5	...	...
57	1 in. to No. 4 (25.0 to 4.75 mm)	...	...	...	...	...	100	95 to 100	...	25 to 60	...	0 to 10	0 to 5	...
67	¾ in. to No. 4 (19.0 to 4.75 mm)	...	...	...	...	...	...	100	90 to 100	...	20 to 55	0 to 10	0 to 5	...
7	½ in. to No. 4 (12.5 to 4.75 mm)	...	...	...	...	...	...	...	100	90 to 100	40 to 70	0 to 15	0 to 5	...
8	⅜ in. to No. 8 (9.5 to 2.36 mm)	...	...	...	...	...	...	...	...	100	85 to 100	10 to 30	0 to 10	0 to 5
3	2 to 1 in. (50 to 25.0 mm)	...	...	...	100	90 to 100	35 to 70	0 to 15	0 to 5	...	...	...	...	...
4	1½ to ¾ in. (37.5 to 19.0 mm)	...	...	...	...	100	90 to 100	20 to 55	0 to 15	0 to 5	...	...	...	...

(b) For exposed concrete, aggregates shall not contain any particles that are deleteriously reactive with alkalies.

FIG. 7–10. continued

(c) Deleterious substances (see map for weathering regions).

Class Designation	Type of Location of Concrete Construction	Maximum allowable (%)				Maximum allowable (%)		
		Clay Lumps and Friable Particles	Chert (Less than 2.40 sp gr SSD)	Sum of Clay Lumps, Friable Particles and Chert (Less Than 2.40 sp gr SSD)	Material Finer Than No. 200 (75-μm) Sieve	Coal and Lignite	Abrasion	Magnesium Sulphate Soundness (5-cycles)
	Severe Weathering Regions							
1S	Footings, foundations, columns and beams not exposed to the weather, interior floor slabs to be given coverings	10.0	...	...	1.0	1.0	50	...
2S	Interior floors without covering	5.0	...	...	1.0	0.5	50	...
3S	Foundation walls above grade, retaining walls, abutments, piers, girders, and beams exposed to the weather	5.0	5.0	7.0	1.0	0.5	50	18
4S	Pavements, bridge decks, driveways and curbs, walks, patios, garage floors, exposed floors and porches, or water-front structures subject to frequent wetting	3.0	5.0	5.0	1.0	0.5	50	18
5S	Exposed architectural concrete	2.0	3.0	3.0	1.0	0.5	50	18

FIG. 7-10. continued

1M	Footings, foundations, columns, and beams not exposed to the weather, interior floor slabs to be given coverings.	10.0	…	…	1.0	1.0	50	…
2M	Interior floors without covering	5.0	…	…	1.0	0.5	50	…
3M	Foundation walls above grade, retaining walls, abutments, piers, girders, and beams exposed to the weather	5.0	8.0	10.0	1.0	0.5	50	18
4M	Pavements, bridge decks, driveways and curbs, walks, patios, garage floors, exposed floors and porches, or water-front structures subject to frequent wetting	5.0	5.0	7.0	1.0	0.5	50	18
5M	Exposed architectural concrete	3.0	3.0	5.0	1.0	0.5	50	18

Moderate Weathering Regions

Negligible Weathering Regions

1N	Slabs subject to traffic abrasion, bridge decks, floors, sidewalks, pavements	5.0	…	…	1.0	0.5	50	…
2N	All other classes of concrete	10.0	…	…	1.0	1.0	50	…

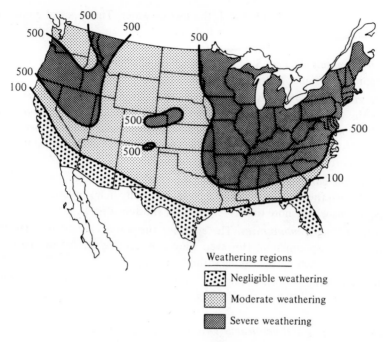

FIG. 7–10. continued

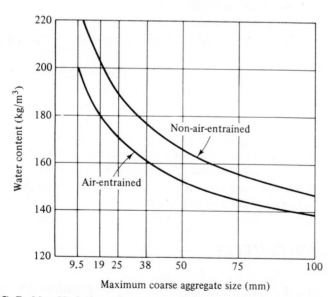

FIG. 7–11. Variation of water requirement with aggregate size for concrete of a given consistency. (Courtesy of the Canadian Portland Cement Association.)

poses. They are usually added with the mixing water. The types of admixtures and their uses can be summarized as follows:

1. *Air-entraining agents.* Such agents entrain air bubbles in the concrete to improve resistance to freezing and thawing.
2. *Accelerating admixtures.* To accelerate the setting and strength development of the concrete, admixtures such as calcium chloride are commonly used. The amount used should not exceed 2% of the mass of the cement in the mix. Use of these admixtures must be carefully controlled because they may cause corrosion or increased drying shrinkage.
3. *Retarding admixtures.* These retard the setting time in the concrete and are used when the concrete sets too rapidly, especially in hot weather or in other situations where more time is required for finishing operations.
4. *Water-reducing admixtures.* These reduce the water content required for a certain slump concrete. However, they may also retard the setting of the concrete or increase the drying shrinkage.
5. *Workability agents.* These improve the workability of the concrete for trowelling operations, when the concrete is to be pumped, or in other special situations.
6. *Dampproofing and permeability-reducing agents.* These may be used to help produce waterproof concrete.
7. *Grouting agents.* Such agents are added to concrete that is to be used for such purposes as grouting or sealing cracks.
8. *Gas-forming agents.* These cause a slight expansion in the concrete prior to hardening.
9. *Superplasticizers.* These are fairly new types of admixtures that make concrete that flows easily without vibration or segregation. As the plasticity of the concrete is increased considerably, less water is required and much higher strengths can be obtained with the same cement content. Concrete has been placed with W/C ratios as low as 0.3 or less. Also, with normal W/C ratios, a more economical mix can be produced as both the water and cement content can be reduced. One problem encountered with the use of superplasticizers in ready-mix concrete is that the increase in plasticity lasts only a short time, usually less than one hour.

7–4 DESIGN CRITERIA

Specifications and design criteria for concrete depend primarily on the strength, exposure, and placing conditions. Detailed proportions can be specified, or general requirements can be given and trial mixes prepared to design a mix meeting the requirements.

7-4.1 The W/C ratio required for concrete depends on exposure conditions and desired strength. If two W/C ratios are determined—one for exposure and one for strength—the lower must be used because they will both be maximum values.

ACI specifications for maximum W/C ratios for concrete in severe exposures are given in table 7–10. CSA specifications are shown in part (a) of table 7–11. Additional CSA requirements for concrete exposed to sulphates are included in their standard A 23.1.

Strength that can be expected from various W/C ratios can be estimated from fig. 7–12 or tables 7–12 and 7–13. Because strength will vary in any concrete design from cylinder to cylinder and from one day's production to the next day's, design strength must be higher than specified strength to ensure that all, or practically all, the concrete produced meets the specified strength. The bands shown for strength in fig. 7–12 give the range of values that can be expected for a given W/C ratio. Using the lower edge of this band to estimate the required W/C ratio will give a conservative value and thus ensure that the average strength will be greater than the specified strength.

Using the tables, it is common practice to use a design strength 3.5 MPa (500 psi) above the specified strength for selection of W/C ratios for the initial trial mixes.

Table 7–10

MAXIMUM WATER/CEMENT RATIOS FOR CONCRETE IN SEVERE EXPOSURES (ACI 211.1–77)*

Type of structure	Structure wet continuously or frequently and exposed to freezing and thawing[a]	Structure exposed to seawater or sulfates
Thin sections (railings, curbs, sills, ledges, ornamental work) and sections with less than 1 in.-cover over steel	0.45	0.40[b]
All other structures	0.50	0.45[b]

* Copyright © 1969, 1973, 1975, 1977, American Concrete Institute, Reprinted from ACI 211.1–77 with the permission of the American Concrete Institute.

[a] Concrete should also be air entrained.

[b] If sulfate resisting cement (Type II or Type V of ASTM C 150) is used, permissible water-cement ratio may be increased by 0.05.

Example 7–3: Using table 7–12, estimate the required W/C ratio for 4000-psi concrete, assuming that the average strength should exceed the specified strength by 500 psi. The concrete is not air entrained.

Table 7–11

CONCRETE QUALITY REQUIREMENTS–CSA

**Maximum Permissible Water-Cement
Ratios–Normal-Density Concrete**

Exposure class	Exposure	W/C ratio by mass, maximum
A	Frequent freeze-thaw when saturated in seawater or subject to deicers	0.45
B	Frequent freeze-thaw when saturated in fresh water or infrequent wetting by seawater Complete, continuous immersion in seawater	0.50
C	Frequent freeze-thaw when not saturated	0.55

**Recommended Range of Total Air Content
Required for Various Exposures**

Class of Exposure	Condition of Exposure	Range of total air content (percentage) required for concretes with indicated size of coarse aggregate			
		10 mm	14 mm	20 mm	40 mm
A	Frequent cycles of freezing in saturated condition and 1. subject to deicing chemicals 2. not subject to deicing chemicals	7 to 10 6 to 9	6 to 9 5 to 8	5 to 8 4 to 7	4 to 7 3 to 6
B	Frequent cycles of freezing and thawing in a saturated condition in fresh water; infrequent wetting by seawater; or complete and continuous immersion in seawater	6 to 9	5 to 8	4 to 7	3 to 6

C	Frequent cycles of freezing and thawing in an un-saturated condition	5 to 8	4 to 7	3 to 6	3 to 6
D*	No exposure to freezing and thawing or the application of deicing chemicals	<5	<4	<3	<3
	Approximate amount of entrapped air in non-air-entrained concrete	3	2.5	2	1

* The use of entrained air is not mandatory for class of exposure D, but under normal conditions it is recommended as an aid to workability and reduced bleeding.

(This information is adapted from CSA Standard CAN 3–A23.1–M 77, Concrete Construction, which is copyrighted by the Canadian Standards Association. Copies may be purchased from the association at 178 Rexdale Boulevard, Rexdale, Ontario, M9W 1R3.)

W/C for 4000 psi is 0.57

W/C for 5000 psi is 0.48

For 4500, try

W/C = 0.52 or 0.53

Example 7–4: Using fig. 7–12, estimate the required W/C ratio for 25-MPa, non-air-entrained concrete, using the lower line of expected values.

The lower line of the strength band for 25 MPa indicates a W/C ratio of approximately 0.60.

Example 7–5: A retaining wall (2-in. cover over reinforcing bars) is exposed to frequent wetting and freezing. If the strength required is 3500 psi, recommend a W/C ratio for the trial mix using ACI specifications.

Strength (4000 psi): W/C required = 0.48 (air-entrained)

Exposure: W/C required = 0.50

∴ Use W/C = 0.48

Example 7–6: A pavement slab is required in a severe climate. If the strength required is 25 MPa, select a W/C ratio, using the CSA criteria.

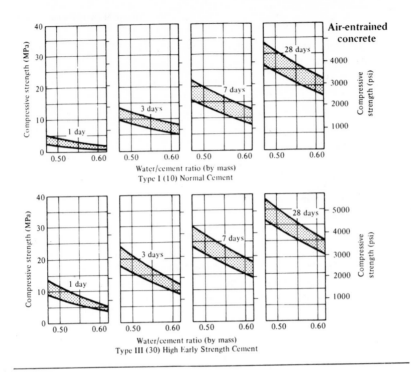

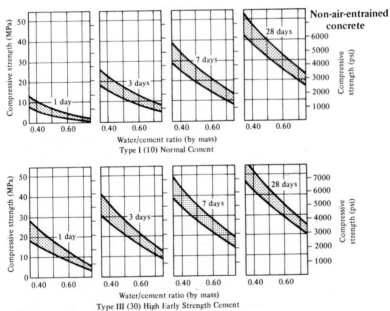

FIG. 7—12. Range of expected strengths for concrete. (Courtesy of the Canadian Portland Cement Association.)

Table 7–12

ESTIMATED AVERAGE STRENGTHS
FOR CONCRETE (psi) (ACI 211.1–77)*

Compressive strength at 28 days (psi)	Water-cement ratio (by weight)	
	Non-air-entrained concrete	Air-entrained concrete
6000	0.41	—
5000	0.48	0.40
4000	0.57	0.48
3000	0.68	0.59
2000	0.82	0.74

* Copyright © 1969, 1973, 1975, 1977, American Concrete Institute. Reprinted from ACI 211.1–77 with the permission of the American Concrete Institute.

Table 7–13

ESTIMATED AVERAGE STRENGTHS FOR CONCRETE (MPa)
(Adapted from Table 7–12)

Compressive Strength, 28 Days (MPa)	Water/Cement Ratio (By Mass)	
	Non-Air-Entrained Concrete	Air-Entrained Concrete
40	0.42	—
35	0.47	0.39
30	0.54	0.45
25	0.61	0.52
20	0.69	0.60
15	0.79	0.71

Strength (25 MPa): W/C = 0.48 (lower line of the chart in fig. 7–12)

Exposure: W/C = 0.45

Use W/C = 0.45

7–4.2 Maximum aggregate size must be specified or chosen. The largest size economically available should be chosen, subject to the following maximum size:

1. One-fifth of the narrowest dimension between the sides of forms;
2. Three-fourths of the minimum clear spacing between the reinforcing bars and forms and between adjacent bars, or
3. One-third of the depth of the unreinforced slabs.

The ACI and the CSA standards use 5 mm, 10 mm, 20 mm, and 40 mm for the common coarse aggregate sizes (No. 4, 3/8 in, 3/4 in, and 1½ in), and therefore these sizes will be used in concrete mix design examples.

Example 7–7: A retaining wall is 200 mm (8 in) thick; the steel has 40 mm (1½ in) of cover. What size aggregate should be used?

$$1/5 \text{ of thickness is } 200/5 = 40 \text{ mm } (8/5 = 1\ 1/2 \text{ in.})$$

$$3/4 \text{ of spacing is } 3/4 \times 40 = 30 \text{ mm } (3/4 \times 1\ 1/2 = 1 \text{ in.})$$

A 25-mm (1-in) aggregate could be used. If the usual aggregate sizes produced in an area are 20 mm (3/4 in) and 40 mm (1 1/2 in), as is often the case, the 20-mm (3/4-in) size would be chosen.

7–4.3 All concrete exposed to freezing and thawing should be air-entrained. Tables 7–11 and 7–14 give the recommended air content for air-entrained concrete based on the degree of exposure, as well as the approximate air content of non-air-entrained concrete.

7–4.4 Design slump depends on the type of structure in which the concrete is to be placed. Slump varies with the water content for any specific mix and, therefore, should be kept as low as possible for purposes of economy. Restricting slump will also reduce the tendency for segregation or bleeding.

However, a certain degree of workability, as measured by slump, is required to ensure that the concrete will fill all the voids in the form without too much vibration.

ACI requirements are given in table 7–15. The CSA standards are similar.

7–5 BASIC TESTS

The basic tests used in mix design and quality control of concrete are those that measure strength, workability (slump), and air content.

7–5.1 The most common strength test uses cylinders 150 mm (6 in) in diameter and 300 mm (12 in) high. These are made in a specified manner, allowed to set,

Table 7–14

AIR CONTENTS FOR CONCRETE (ACI 211.11–77)*

	Nominal Maximum Size of Aggregate in In. (mm)						
	3/8 (10)	1/2 (12.5)	3/4 (20)	1 (25)	1½ (40)	2 (50)	3 (70)
Non-air-entrained concrete— approximate amount of entrapped air (percentage)	3	2.5	2	1.5	1	0.5	0.3
Air-entrained concrete— recommended average total air content (percentage)							
—Mild exposure[a]	4.5	4.0	3.5	3.0	2.5	2.0	1.5
—Moderate exposure[b]	6.0	5.5	5.0	4.5	4.5	4.0	3.5
—Extreme exposure[c]	7.5	7.0	6.0	6.0	5.5	5.0	4.5

* Copyright © 1969, 1973, 1975, 1977, American Concrete Institute. Reprinted from ACI 211.1–77 with the permission of the American Concrete Institute.

[a] Mild exposure—not exposed to freezing.

[b] Moderate exposure—exposed to freezing in an unsatured condition and not exposed to de-icing chemicals.

[c] Extreme exposure—exposed to freezing in a saturated condition or to de-icing chemicals.

Table 7–15

RECOMMENDED SLUMPS (ACI 211.1–77)*

	Slump (in.)		Slump (cm)	
Types of Construction	Maximum[a]	Minimum	Maximum	Minimum
Reinforced foundation walls and footings	3	1	8	2
Plain footings, caissons, and substructure walls	3	1	8	2
Beams and reinforced walls	4	1	10	2
Building columns	4	1	10	2
Pavements and slabs	3	1	8	2
Mass concrete	2	1	5	2

* Copyright © 1969, 1973, 1975, 1977, American Concrete Institute. Reprinted from ACI 211.1–77 with permission of the American Concrete Institute.

[a] May be increased 1 in. (2 cm) for methods of consolidation other than vibration.

and then cured in the laboratory in a humid room until a specified age. Usually the strength is determined 28 days after casting because this is the design strength or common specified strength. However, 7-day strength may also be obtained either as an indication of the expected 28-day strength or as a specified strength.

Cylinders are capped with a sulphur compound to provide flat, smooth ends that are perpendicular to the axis of the sample. They are then tested to failure in a compression machine. Strength is the total load divided by the cross-sectional area.

Example 7–8: The total load carried by a cylinder is 462 KN.

Area (using a 150-mm diameter mold) = 0.01767 m^2

$$\text{Compressive strength} = \frac{462 \text{ KN}}{0.01767 \text{ m}^2} = 26{,}100 \text{ KN/m}^2$$

$$= 26.1 \text{ MPa}$$

Example 7–9 Failure load on a standard 6-inch cylinder is 101,000 lb.

$$\text{Compressive strength} = \frac{101{,}000 \text{ lb}}{\pi \times 3^2 \text{ in}^2} = 3570 \text{ psi}$$

The necessity for waiting 28 days before finding out if a concrete meets the specifications is inconvenient, to say the least. A structure could already be poured to a level far above before tests throw suspicion on the quality of concrete in the foundation. Some authorities allow acceleration of curing to obtain 28-day strength measurements earlier. Two methods to do this are (1) to boil the cylinder a certain period of time or (2) to cure the cylinder in an autogenous curing box that is insulated, allowing curing to accelerate due to the rise in temperature from hydration. In both methods the cylinder can be tested at 2 days of age and the 28-day strength estimated very reliably. Calibration tests to give the relationship between the 2-day strength and the 28-day strength usually must be conducted for each mixture.

When concrete is to be subjected to bending loads, such as in pavement slabs, a flexural strength is often specified. A concrete beam 150 mm × 150 mm (6 in × 6 in) and 900 mm (36 in) long is cast. Usually the beam is loaded at the third points (fig. 7–13). The flexural strength, or modulus of rupture, is obtained by multiplying the failure load by l/wd^2, where l, w, and d are the length (between supports), width, and depth of the beam, respectively.

Example 7–10 The failure load for a 150 mm × 150 mm beam, with a span length of 450 mm, is 31.6 kN. Find its flexural strength.

$$31.6 \text{ kN} \times \frac{0.45 \text{ m}}{0.15 \text{ m} \times 0.15^2 \text{ m}^2} = 4.2 \text{ MPa}$$

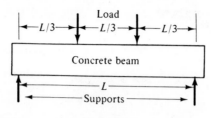

FIG. 7–13.

7–5.2 The main tests performed on plastic concrete are the slump test, and, for air-entrained concrete, the air content test.

In the slump test the concrete is placed in a cone 300 mm (12 in) high and consolidated by rodding. The cone is lifted up, and the amount that the concrete slumps or drops from the original 300-mm height is recorded. This is the primary test to measure the consistency or workability of the material.

The slump of ordinary structural concrete is usually 50-100 mm (2-4 in). Concrete with slump values of 100-150 mm (4-6 in) would be classified as high slump concrete. Highway pavements may require "zero" slump or very low slump (0-30 mm or 0-1 in) concrete.

7–5.3 Air content is measured by a volumetric method or a pressure method. In both methods a bowl of known volume is filled, and the top part of the apparatus is clamped on. With the volume apparatus a standpipe is filled with water, and the apparatus is inverted and agitated until the concrete pulls away from the base. Then it is rolled until no further drop in water level occurs. The drop in the water level is calibrated to give air content as a percentage.

In the pressure method an air chamber on top of the apparatus is pressurized to a certain level. A valve is opened between this chamber and the concrete. Since the pressure multiplied by the volume remains constant, the drop in air pressure indicates the percentage of air. (The pressure method should not be used for lightweight or porous aggregates.)

Figures 7–14 and 7–15 show a slump test and an air meter.

7–6 CONCRETE MIX DESIGN

Concrete mix design involves:

1. Selecting the water/cement ratio, slump, aggregate size, and air content, based on the strength requirement, exposure conditions, and placing conditions. These may be specified, or the required strength and other conditions to be met may be specified and the mix designer allowed to select the other criteria.

FIG. 7—14.
Slump test.

FIG. 7—15. Pressure-type air meter.

2. Evaluating and accepting the aggregates and other materials.

3. Conducting trial mixes to find the optimum proportions of fine aggregate, coarse aggregate, cement, and water to provide the required workability and properties with the most economical combination of materials.

4. Calculating yields and quantities required for production.

Selection of the basic criteria has been discussed in section 7—4. The tests required to evaluate the aggregates have been outlined in chapter 4 and section 7—3.

Two methods used to prepare and evaluate trial mixes are outlined here. The water/cement ratio method follows the basic procedure outlined in "Design and Control of Concrete Mixtures" from the Canadian Portland Cement Association. The ACI method is described in detail in its Standard ACI 211.1—77.

Concrete mix proportions are usually given in units of kg/m^3 or lb/yd^3. (1 kg/m^3 = 1.6855 lb/yd^3 or 1 lb/yd^3 = 0.5933 kg/m^3.)

7—6.1 In the water/cement ratio method of proportioning, the water and cement for the desired size of mix are measured out and mixed, and then the aggregates are added until the desired workability is reached with a mix of acceptable finishing properties that is not too harsh or over-sanded.

Figure 7—16, giving the estimated mass of aggregate required per unit mass of water, and table 7—16, listing percentages of fine aggregate, will help in measuring out the initial amounts required.

Example 7—11 illustrates the preparation of a trial mix using the water/cement ratio method.

Example 7—11. Criteria: W/C, 0.50; maximum aggregate size, 20 mm (3/4 in); fineness modulus of the sand, 2.70; non-air-entrained; desired slump, 75–100 mm.

1. Choose the amount of cement required [about 3 kg of cement will usually provide enough mix for one standard 150 mm $\times$ 300 mm (6 in $\times$ 12 in) cylinder.] Assuming that three cylinders are required, use 10 kg of cement.
2. Determine the amount of water required.

$$10 \text{ kg} \times 0.50 = 5 \text{ kg}$$

3. Estimate the mass of aggregate required (see fig. 7—16).

$$\text{Total mass of aggregate} = 8.5 \times 5 = 42.5 \text{ kg}$$

4. Obtain the mass of fine aggregate and coarse aggregate (see table 7—16).

$$\text{Fine aggregate:} \quad 0.40 \times 42.5 = 17.0 \text{ kg}$$

$$\text{Coarse aggregate:} \quad 0.60 \times 42.5 = 25.5 \text{ kg}$$

Weigh out about 20% more fine and coarse aggregates to allow for adjustments to provide the characteristics desired of the mix.

5. Mix the water and the cement together, and add fine and coarse aggregates until the workability and the trowelling characteristics are suitable. Measure workability with a slump test and inspect the mixture for harshness. That is, is it "harsh," "over-sanded," or "good"? (See fig. 7—17.) (If air-entrained concrete were being designed, an air content test would also be required.)

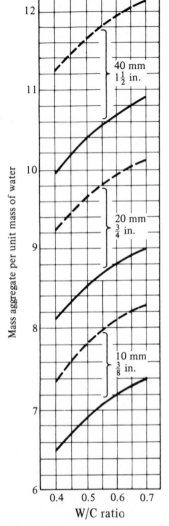

——————— Non-air-entrained concrete

– – – – – Air-entrained concrete

Note: For low slump concrete [25–50 mm (1–2 in.)]
       increase aggregates by 8%.
       For high slump concrete [150–175 mm (6–7 in.)]
       decrease aggregates by 6%.

FIG. 7–16. Approximate mass of aggregate per unit mass of water for concrete of medium consistency [slump: 75–100 mm (3–4 in.)].

6. Find the density of the mixture by obtaining the mass of the material in a container of known volume.

7. Cast cylinders for strength tests.

8. Calculate the mix proportions per m^3. Assume that 17.8 kg of fine aggregate and 25.2 kg of coarse aggregate were used to provide an acceptable mix, and that the mass in a 0.010-m^3 container was 23.1 kg.

$$\text{Density} = \frac{23.1 \text{ kg}}{0.01 \text{ m}^3} = 2310 \text{ kg/m}^3$$

Table 7–16

FINE AGGREGATE AS A PERCENTAGE
OF TOTAL AGGREGATE

W/C	Max. Agg. Size (in.)	(mm)	Fineness Modulus 2.50	2.70	2.90
0.40	3/8	10	50%	52%	54%
	3/4	20	35	37	39
	1½	40	29	31	33
0.50	3/8	10	53	55	57
	3/4	20	38	40	42
	1½	40	32	34	36
0.60	3/8	10	54	56	58
	3/4	20	40	42	44
	1½	40	33	35	37
0.70	3/8	10	55	57	59
	3/4	20	41	43	45
	1½	40	34	36	38

(a)   (b)   (c)

FIG. 7–17. Harshness of concrete: (a) *harsh mix*—not enough cement-sand mortar to fill the spaces between the coarse aggregate particles; (b) *good mix*—with a light troweling all the spaces between the coarse particles are filled (note the good supply of coarse particles at the edge of the pile); (c) *over-sanded mix*—finishes well but with an excess of mortar, making it an uneconomical and possibly porous mix.

Mass used:

$$\text{Cement} = 10.0 \text{ kg}$$
$$\text{Water} = 5.0 \text{ kg}$$
$$\text{Fine aggregate} = 17.8 \text{ kg}$$
$$\text{Coarse aggregate} = \underline{25.2 \text{ kg}}$$
$$Total = 58.0 \text{ kg}$$

Mass per m³:

$$\text{Cement} = \frac{10.0}{58.0} \times 2310 = 398 \text{ kg/m}^3$$

$$\text{Water} = \frac{5.0}{58.0} \times 2310 = 199 \text{ kg/m}^3$$

$$\text{Fine aggregate} = \frac{17.8}{58.0} \times 2310 = 709 \text{ kg/m}^3$$

$$\text{Coarse aggregate} = \frac{25.2}{58.0} \times 2310 = 1004 \text{ kg/m}^3$$

The cement factor is 398 kg/m³, and the percentage of fine aggregate is 17.8/
(17.8 + 25.2) = 41%.

A series of tests is often conducted with different proportions of aggregates to obtain the most economical combination. Table 7–17 summarizes the results of such a series of tests.

7–6.2 In the ACI method, the estimated proportions per cubic yard or cubic metre are obtained, and from these values the quantities for a trial mix are

Table 7–17

SAMPLE RESULTS OF TRIAL MIXES*

Batch no.	Slump (mm)	Air content (%)	Density (kg/m³)	Cement content (kg/m³)	Fine aggregate (% of total aggregate)	Workability
1	80	5.4	2306	320	34.0	Excellent
2	70	4.9	2312	329	27.4	Harsh
3	60	5.1	2310	326	35.5	Excellent
4	80	4.7	2323	320	30.5	Good

* Courtesy of the Canadian Portland Cement Association.

found. The dry materials are mixed and water added as required to give the correct slump.

Criteria for the mix are either obtained from the project specifications or selected according to the requirements given in section 7-4.

The method used to make the trial mix is illustrated in example 7-12.

Example 7-12: Mix criteria and materials: W/C, 0.50; maximum aggregate size, 20 mm (3/4 in); slump, 75-100 mm (3-4 in); non-air-entrained. Aggregates: coarse, RD_{SSD} = 2.66; dry rodded density, 1540 kg/m^3 (96 lb/ft^3); fine, RD_{SSD} = 2.64; fineness modulus, 2.80.

1. Estimate the amount of water required. From table 7-18 the water requirement is 340 lb/yd^3.

2. Determine the amount of cement required.

$$340/0.50 = 680 \text{ lb/yd}^3$$

3. Find the amount of coarse aggregate required. From table 7-19 this is 0.62 yd^3. The mass of the coarse aggregate is

$$0.62 \times 96 \text{ lb/ft}^3 \times 27 \text{ ft}^3/\text{yd}^3 = 1607 \text{ lb}$$

4. Calculate the amount of fine aggregate required. The volume (see section 7-6.4 for details of the volume calculations) occupied by the water, cement, coarse aggregate, and air are:

$$\text{Water: } 340/62.4 = 5.45 \text{ ft}^3$$
$$\text{Cement: } 680/(3.15 \times 62.4) = 3.46 \text{ ft}^3$$
$$\text{Coarse aggregate: } 1607/(2.66 \times 62.4) = 9.68 \text{ ft}^3$$
$$\text{Air (see table 7-14): 2\% of 27} = \underline{0.54 \text{ ft}^3}$$
$$\textit{Total} = 19.13 \text{ ft}^3$$

Therefore the volume of the fine aggregate required is 27 - 19.13 = 7.87 ft^3. The mass needed is 7.87 $\times$ 2.64 $\times$ 62.4 = 1296 lb. [An alternative method of finding the amount of fine aggregate is to subtract the total mass of the water, cement, and coarse aggregate from the estimated mass of the concrete given in table 7-20. In this example, the mass of sand required would be 3960 - (340 + 680 + 1607) = 1333 lb.]

5. Measure the quantities needed for a trial mix. Assuming that a mix batch of 0.75 ft^3 is to be made, these quantities are:

Table 7-18

APPROXIMATE MIXING WATER REQUIREMENTS
(ACI 211.1-77)*

(a) lb/yd^3

Slump (in.)	Water (lb per yd^3 of concrete for indicated nominal maximum sizes of aggregate)						
	3/8 in.	1/2 in.	3/4 in.	1 in.	1 1/2 in.	2 in.	3 in.
Non-air-entrained concrete							
1 to 2	350	335	315	300	275	260	240
3 to 4	385	365	340	325	300	285	265
6 to 7	410	385	360	340	315	300	285
Air-entrained concrete							
1 to 2	305	295	280	270	250	240	225
3 to 4	340	325	305	295	275	265	250
6 to 7	365	345	325	310	290	280	270

(b) kg/m^3

Slump (cm)	Water (kg/m^3 of concrete for indicated maximum sizes of aggregate in mm)						
	10	12.5	20	25	40	50	70
Non-air-entrained concrete							
3 to 5	205	200	185	180	160	155	145
8 to 10	225	215	200	195	175	170	160
15 to 18	240	230	210	205	185	180	170
Air-entrained concrete							
3 to 5	180	175	165	160	145	140	135
8 to 10	200	190	180	175	160	155	150
15 to 18	215	205	190	185	170	165	160

Note: These quantities of mixing water are for use in computing cement factors for trial batches. They are maxima for reasonably well-shaped angular coarse aggregates graded within limits of accepted specifications.

Water: 340 × 0.75/27 = 9.44 lb or   4.28 kg

Cement: 680 × 0.75/27 = 18.89 lb or   8.57 kg

Fine aggregate: 1296 × 0.75/27 = 36.00 lb or 16.33 kg

Coarse aggregate: 1607 × 0.75/27 = 44.64 lb or 20.25 kg

Total = 3923 lb   108.97 lb or 49.43 kg

6. Mix the cement, aggregates, and part of the water together. Add more water, mixing each time water is added, until the desired slump is reached. The amount of water required may be more or less than 4.28 kg. Observe the mix for proper workability and finishing qualities.

7. Make a slump and an air test (if mixing air-entrained concrete).

8. Determine the density by measuring the mass of concrete in a container of known volume. (*Note:* If the proportions were established by the volume method in step 4, and the estimated water and air contents coincided with the actual amounts in the mix, the density should be 3923 lb/yd^3 or 145.3 lb/ft^3.)

9. Cast the samples required for strength testing.

Table 7–19

VOLUME OF COARSE AGGREGATE PER UNIT
VOLUME OF CONCRETE (ACI 211.1–77)

Maximum size of aggregate		Volume of dry-rodded coarse aggregate per unit volume of concrete for different fineness moduli of sand			
in.	mm	2.40	2.60	2.80	3.00
3/8	10	0.50	0.48	0.46	0.44
1/2	12.5	0.59	0.57	0.55	0.53
3/4	20	0.66	0.64	0.62	0.60
1	25	0.71	0.69	0.67	0.65
1-1/2	40	0.75	0.73	0.71	0.69
2	50	0.78	0.76	0.74	0.72
3	70	0.82	0.80	0.78	0.76

* Copyright © 1969, 1973, 1975, 1977, American Concrete Institute. Reprinted from ACI 211.1–77 with the permission of the American Concrete Institute.

Note: These volumes are selected to produce concrete with a degree of workability suitable for usual reinforced construction. For less workable concrete such as required for concrete pavement construction they may be increased about 10%.

Table 7−20

ESTIMATE OF MASS OF CONCRETE
(ACI 211.1−77)*

(a) lb/yd³

Maximum size of aggregate (in.)	First estimate of concrete weight (lb per yd³)	
	Non-air-entrained concrete	Air-entrained concrete
3/8	3840	3690
1/2	3890	3760
3/4	3960	3840
1	4010	3900
1 1/2	4070	3960
2	4120	4000
3	4160	4040

(b) kg/m³

Maximum size of aggregate (mm)	First estimate of concrete weight (kg/m³)	
	Non-air-entrained concrete	Air-entrained concrete
10	2285	2190
12.5	2315	2235
20	2355	2280
25	2375	2315
40	2420	2355
50	2445	2375
70	2465	2400

10. Revise the estimated quantities per cubic yard established in step 4. If 4.53 kg of water is required to produce the desired slump, the revised amount of water is (4.53/4.28) × 340 = 360 lb/yd³. The cement required will be 360/0.50 = 720 lb/yd³. The quantities of fine and coarse aggregates may also be altered to improve the workability and finishing qualities of the mix.

11. Results of a series of trial mixes are then summarized in a table such as that shown in table 7–17 and the design mix proportions established.

7–6.3 Aggregates should be in the saturated, surface-dry condition for the trial mix. If this is not possible, adjustments have to be made to the quantities and results.

Example 7–13: Assume same data as for example 7–11 except that (1) the sand is dry and has an absorption of 0.8% and (2) the coarse aggregate is wet, with a water content of 3.8% and an absorption of 1.2% (or an excess water content of 2.6%).

Complete steps 1, 2, and 3 as in example 7–11.

4. Estimated fine aggregate = 17.0 kg

$$\therefore \text{ extra water required} = 0.008 \times 17.0 = 0.14 \text{ kg.}$$

Estimated coarse aggregate = 25.5 kg

$$\therefore \text{ excess water} = 0.026 \times 25.5 = 0.66 \text{ kg.}$$

Therefore, the water required to be added to mix is

$$5.0 \text{ kg} + 0.14 \text{ kg} - 0.66 \text{ kg} = 4.48 \text{ kg.}$$

Complete steps 5, 6, and 7 as in example 7–11.

8. Assume that 18.5 kg of the dry fine aggregate and 24.2 kg of the wet coarse aggregate, as well as the 4.48 kg of water, are used.

Quantities used:

$$\text{Cement} = 10.0 \text{ kg}$$
$$\text{Water} = 4.48 \text{ kg}$$
$$\text{Fine aggregate} = 18.5 \text{ kg}$$
$$\text{Coarse aggregate} = 24.2 \text{ kg}$$

However, the fine aggregate required

$$0.008 \times 18.5 = 0.15 \text{ kg extra water.}$$

The coarse aggregate contained

$$0.026 \times 24.2 = 0.63 \text{ kg excess water.}$$

Therefore, the actual quantities (with aggregates in the *SSD* condition) are:

$$\text{Cement} = 10.0 \text{ kg}$$

$$\text{Water} = 4.48 - 0.15 + 0.63 = 4.96 \text{ kg}$$

$$\text{Fine aggregate} = 18.5 + .15 = 18.65 \text{ kg}$$

$$\text{Coarse aggregate} = 24.2 - 0.63 = 23.57 \text{ kg}$$

These values should be used for the balance of the calculations in example 7–11. Note that the W/C ratio is not quite 0.50 (4.96/10 = 0.496). For subsequent trial mixes the estimated aggregates will be much closer to the amounts used and therefore the difference in W/C ratios will be negligible.

7–6.4 Quantities required per m^3 (or yd^3) can be obtained by calculating absolute volumes. This procedure may be used to check the results of the density test conducted in a trial mix or to adjust proportions obtained by some other method.

Example 7–14: Approximate quantities for 1 m^3 are

$$\text{Cement} = \quad 300 \text{ kg}$$

$$\text{Water} = \quad 178 \text{ kg}$$

$$\text{Fine aggregate} = \quad 670 \text{ kg}$$

$$\text{Coarse aggregate} = 1040 \text{ kg}$$

The air content is 5.0%.

The relative densities in saturated, surface-dry conditions are 2.64 and 2.68 for fine and coarse aggregates respectively. (The RD for portland cement is 3.15.)

The volume can be calculated using equation (1–2)

$$\text{Cement} = \frac{300}{3.15 \times 1000} = 0.095 \text{ m}^3$$

$$\text{Water} = \frac{178}{1.0 \times 1000} = 0.178 \text{ m}^3$$

$$\text{Fine aggregate} = \frac{670}{2.64 \times 1000} = 0.254 \text{ m}^3$$

$$\text{Coarse aggregate} = \frac{1040}{2.68 \times 1000} = 0.388 \text{ m}^3$$

$$\text{Air} = 0.05 \times 1 \text{ m}^3 = 0.050 \text{ m}^3$$

$$\text{Total} = 0.692 \text{ m}^3$$

Because this does not equal 1 m³, the adjusted quantities per m³ can be obtained by multiplying the original masses by 1/0.965; for example, cement per m³ = 300 × (1/0.965) = 311 kg.

The final results are as follows:

	Approx mass per m³ (kg)	Volume (m³)	Adjusted mass (kg)
Cement	300	0.095	311
Water	178	0.178	184
Fine aggregate	670	0.254	694
Coarse aggregate	1040	0.388	1078
Air (5%)		0.050	
Total	2188	0.965	2267

Note: Another calculation of absolute volume will not total exactly 1 m³ because the air content was not adjusted, but the difference will be negligible.

Example 7–15: Estimated quantities for 1 yd³ of concrete are:

Cement	615 lb
Water	340 lb
Fine aggregate	1260 lb
Coarse aggregate	1740 lb

The air content is 2% (non-air-entrained concrete, entrapped air only). The specific gravity values (*SSD* condition) are 2.66 and 2.70, respectively, for the fine and the coarse aggregates. Check the yield.

Volumes, in ft³, are:

$$\text{Cement} \qquad \frac{615}{3.15 \times 62.4} = 3.13 \text{ ft}^3$$

$$\text{Water} \qquad \frac{340}{62.4} = 5.45 \text{ ft}^3$$

$$\text{Fine aggregate} \qquad \frac{1260}{2.66 \times 62.4} = 7.59 \text{ ft}^3$$

$$\text{Coarse aggregate} \qquad \frac{1740}{2.70 \times 62.4} = 10.33 \text{ ft}^3$$

$$\text{Air} \qquad 0.02 \times 27 = \underline{0.54 \text{ ft}^3}$$

$$\text{Total} \qquad \qquad 27.04 \text{ ft}^3$$

This result is within 0.2% of estimated volume of 1 yd^3, and further adjustments would not be justified.

7–7 QUALITY CONTROL OF CONCRETE

After the mix design for a concrete project has been established, the quality of the concrete produced must be checked to ensure that the required characteristics of the material produced meet the specifications. This involves tests to check:

— materials
— proportioning
— plastic concrete
— quality of hardened concrete

7–7.1 The aggregates that are used are tested periodically to ensure that their properties have not changed greatly from those of the materials tested for the mix design. The sand gradation is especially important, and the usual requirement is that the fineness modulus not vary by more than 0.20 from the original value.

7–7.2 The proportioning of the materials must be changed whenever the moisture content of the aggregate changes significantly. The mix design is based on aggregates in the saturated, surface-dry condition, and the aggregate may be wet, containing extra water, or not fully saturated.

The Speedy moisture tester (fig. 7–18) is often used to find the water content of sand. A known mass of sand, measured on a small balance, is placed

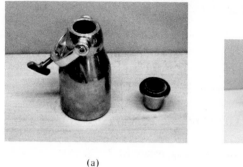

(a)   (b)

FIG. 7–18. Speedy moisture tester. (Courtesy of Humbolt Manufacturing Co.)

in a pressure cell. A reagent is added to the other part of the cell. When the two parts are joined and the cell shaken, the water and the reagent produce a gas. The pressure resulting from this gas varies with the amount of gas produced and, therefore, the water content of the sand. The pressure gauge is calibrated to give the percentage of water in the sand.

Electric moisture meters are also often used to obtain water content, as are various methods of quick drying.

Proportions are corrected as shown in the following example:

Example 7–16: The mix design is

$$\text{Cement} = 285 \text{ kg/m}^3 \ (480 \text{ lb/yd}^3)$$

$$\text{Water} = 157 \text{ kg/m}^3 \ (265 \text{ lb/yd}^3)$$

$$\text{Fine aggregate} = 682 \text{ kg/m}^3 \ (1150 \text{ lb/yd}^3)$$

$$\text{Coarse aggregate} = 1180 \text{ kg/m}^3 \ (1990 \text{ lb/yd}^3)$$

$$\text{Air} = 5\%$$

Concrete is mixed in 3.8-m^3 (5-yd^3) batches. The sand is wet: it contains 4.2% water; absorption is 0.6%. The coarse aggregate is almost dry: it contains 0.5% water; absorption is 1.3%. Find the batch proportions. (*Note*–calculations in pounds based on 5.00 yd^3 batch.)

1. The mass required per batch by the design is:

$$\text{Cement} = 285 \times 3.8 = 1083 \text{ kg} \quad (2400 \text{ lb})$$

$$\text{Water} = 157 \times 3.8 = \ 597 \text{ kg} \quad (1325 \text{ lb})$$

$$\text{Fine aggregate} = 682 \times 3.8 = 2592 \text{ kg} \quad (5750 \text{ lb})$$

$$\text{Coarse aggregate} = 1180 \times 3.8 = \underline{4484 \text{ kg}} \quad \underline{(9950 \text{ lb})}$$

$$Total = 8756 \text{ kg } (19{,}425 \text{ lb})$$

2. Excess water in the fine aggregate is:

$$(4.2 - 0.6)\% \text{ of } 2592 = 93 \text{ kg } (207 \text{ lb})$$

Extra water required for the coarse aggregate is:

$$(1.3 - 0.5)\% \text{ of } 4484 = 36 \text{ kg } (80 \text{ lb})$$

3. Therefore, the mass required per batch is:

$$\text{Cement} = 1083 \text{ kg} \quad (2400 \text{ lb})$$

$$\text{Water} = 597 - 93 + 36 = \quad 540 \text{ kg} \quad (1198 \text{ lb})$$

$$\text{Fine aggregate} = 2592 + 93 = 2685 \text{ kg} \quad (5957 \text{ lb})$$

$$\text{Coarse aggregate} = 4484 - 36 = \underline{4448 \text{ kg}} \quad \underline{(9870 \text{ lb})}$$

$$Total = 8756 \text{ kg } (19{,}425 \text{ lb})$$

(*Note:* The total mass per batch does not change.)

Specifications usually require that the materials be measured in batches, within the following degrees of accuracy:

Cement	± 1%
Water	± 2%
Aggregates	± 1%
Admixtures	± 3%

7–7.3 ASTM Standard C94, standard specification for Ready Mixed Concrete, requires one strength test, consisting of two cylinders, for each 150 yd^3 of concrete produced, with a minimum of one test for each class of material each day. The strength is the average of the test results from the two cylinders. This strength should meet the following requirements.

1. For concrete in structures designed by the ultimate strength method or for prestressed concrete, not more than 10% of the tests should have values less than the specified strength and the average of any three consecutive tests should be equal to or greater than the specified strength.

2. For other concrete, not more than 20% of the tests should have values less than the specified strength and the average of any six consecutive tests must be as specified.

This standard also requires that a slump test and an air content test be made at least with every strength test, more often if necessary.

Canadian specifications are similar, one strength test for each 100 m³ of concrete, with the requirements that the average of all sets of three consecutive tests meet the specifications, and that no one test result fail by more than 3.5 MPa.

7–7.4 The standard deviation value is used in quality control and in designating a design strength to produce concrete that will adequately meet the specified strength. It is calculated as follows:

$$SD = \sqrt{\frac{\Sigma(x - \bar{x})^2}{N}}$$

> *where* SD is the standard deviation
> x is the test strength
> $\bar{x}$ is the mean strength
> N is the number of tests

Example 7–17. The results of 12 strength tests are 23.8, 24.3, 20.7, 26.2, 24.1, 23.4, 26.8, 22.7, 19.4, 23.4, 21.5, and 19.5 MPa. Find the standard deviation.

x	$(x - \bar{x})^2$
23.8	0.67
24.3	1.74
20.7	5.20
26.2	10.37
24.1	1.25
23.4	0.18
26.8	14.59
22.7	0.08
19.4	12.82
23.4	0.18
21.5	2.19
19.5	12.11
275.8	61.38

$$\bar{x} = \frac{275.8}{12} \qquad SD = \sqrt{\frac{61.38}{12}} = 2.26 \text{ MPa}$$

$$= 22.98$$

A standard deviation of 2.26 MPa (330 psi) would indicate very consistent concrete. Values of up to 3.5 MPa (500 psi) are considered very good.

The standard deviation may also help in the assigning of strengths for trial mixes that can be used to ensure that the end product meets the minimum specification requirements satisfactorily. A typical requirement is that, for consistent concrete (SD < 3.5 MPa or 500 psi), the average strength for the design mix be equal to the specified minimum required strength plus 1.4 times the SD value.

7–7.5   Temperature measurements are also commonly taken in the fresh, plastic concrete, especially on hot or cold days when special precaution must be taken.

Various types of equipment are now available to allow the inspector to check the cement content and aggregate gradation using a sample of fresh concrete.

7–7.6   Many other tests can be conducted to check the strength of concrete at various stages. New methods are also making their appearance. Lab-cured cylinders actually indicate only the potential strength of the concrete, and the actual strength in place may be quite different, depending on construction conditions.

Operation of the maturity meter (fig. 7–19) is based on the fact that strength varies with the amount of hydration that has taken place. The amount of hydration can be estimated by the amount of heat produced by the concrete, because heat is also a product of the hydration process. A probe, coated with grease, is inserted in the concrete when it is poured. The cumulative degree hours of heat given off is recorded, and this value can indicate concrete strength. A calibration test should usually be conducted to establish the relationship between the heat of hydration and the strength for each mixture.

Another type of test involves casting small rods with enlarged ends in the concrete. These are pulled out with a hand-operated hydraulic jack. The force required to pull them out is directly related to the strength of the concrete.

Concrete that has been poured but becomes suspect later due to poor cylinder test results can be checked by taking cores out with diamond-tipped core barrels.

The rebound hammer (fig. 7–20) tests the surface hardness of concrete. It measures the rebound of a spring-loaded plunger after it has hit the concrete surface.

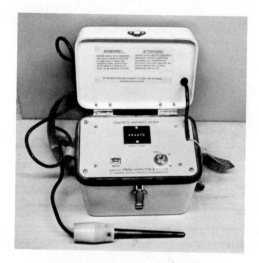

FIG. 7–19. Maturity meter.

FIG. 7–20. Rebound hammer.

A power-activated gun that fires a hardened probe in the concrete can also be used to check strength.

Other types of tests using sonar waves, nuclear rays, X-rays, electrical resistivity, and other methods are being researched as means to measure quality of concrete.

7–8 CURING AND CONSTRUCTION PRACTICES

Concrete can be proportioned and mixed to acceptable standards, but the construction practices followed in placing and curing the materials also have to be satisfactory to ensure the quality of the finished product.

7–8.1 Curing is one of the most important aspects of concrete construction. Poor curing and the addition of extra water at the site are probably the two most common factors that contribute to poor quality concrete.

Proper curing requires (1) water and (2) a favorable temperature. When concrete is allowed to dry, the process of hydration ceases because water is required to react with the cement. (See fig. 7–2.) Concrete will gain strength indefinitely (although at a slower rate) if kept moist. If dried, it is difficult to get it resaturated, and even if it is resaturated, the ultimate strength will probably still be less than if it had never dried out.

Excessive evaporation of water from the surface causes shrinkage of the surface concrete and may lead to cracks.

Low temperatures slow the rate of hydration. The rate of gain of strength is slower at temperatures below 10°C (50°F) and is greatly retarded at 5°C (41°F). At freezing, little or no hydration occurs. Of course, freezing also causes expansion of the water in concrete and leads to disruptive stresses if it occurs before the concrete is strong enough to resist them.

Methods of curing used to ensure that moisture is present include:

— ponding
— sprinkling or fogging
— wet coverings, such as burlap, kept sprayed
— waterproof paper
— plastic film
— curing compounds that form a membrane when sprayed on the surface
— steam curing (see Section 7–8.2).

The duration of the curing period varies according to the type of project and the type of concrete. A period of 7 days is often specified.

7–8.2 The hydration of concrete is often accelerated. This shortens the time required for curing and protection, especially important for precast concrete when forms have to be reused each day for efficient operation, and for winter concreting when the cost of protecting concrete from freezing is high.

Acceleration methods used include steam curing, using high early strength cement (Type 30 or III) and using accelerating admixtures.

Steam curing is often used in precast plants. Steam is applied about 4-5 hours after pouring. This raises the temperature as well as ensuring that water is present. The steam is usually shut off in time to allow the concrete to cool and be removed within 24 hours of casting. Figure 7–21 shows the amount of strength gain that is possible using steam.

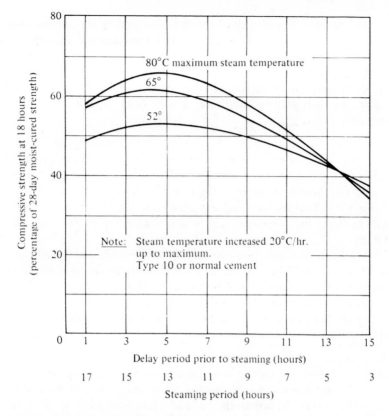

FIG. 7–21. Relationship between strength after 18 hours and conditions of steam curing. (Courtesy of the Canadian Portland Cement Association.)

The use of high early strength cement also allows the period of curing to be reduced significantly, often to 3 days or less. This cement will usually reach over 80% of design strength in 3 days.

Accelerating agents speed up rate of hydration and, therefore, strength gain. Calcium chloride is often used; the amount should not exceed 2% of the mass of the cement, and it cannot be used in reinforced concrete as it is a corroding agent.

7–8.3 ASTM C94 gives specifications for the mixing and transporting of concrete.

The time required for mixing in stationary mixers is 1 minute for the first cubic yard, and 15 seconds for each additional cubic yard or fraction. (One minute for the first cubic metre, and 20 seconds for each additional cubic metre or fraction).

Mixing time in ready mixed trucks is governed by the number of revolutions of the mixing drum. Generally 70-100 revolutions at the mixing speed (6-18 rpm) are required. All revolutions over 100 must be made at the agitating speed (2-6 rpm).

Concrete must be discharged from ready mixed trucks within 1½ hours or 300 drum revolutions (whichever comes first) according to the ASTM requirement.

7–8.4   The concrete must be placed in the forms so that no voids are left and there is no segregation in the mix. Buckets, chutes, pumps, and belt conveyors are used to move the concrete to the forms. It should be deposited as close as possible to its final location and not allowed to fall freely for too great a distance. In forms the concrete should be deposited in layers 200-500 mm (8-20 in) thick.

Vibration is required to consolidate the concrete and fill all the voids. The vibrators should be inserted into the concrete about every 50 cm (18 in) along the form and should penetrate the layer below. Overvibration (about 15 s) may cause segregation or reduce the amount of entrained air in the concrete.

Joints between pours must be dampened and covered with a layer of mortar to ensure a tight, waterproof joint.

7–8.5   Joints must be used in concrete construction. Improper joint construction is a frequent cause of cracking and, therefore, of at least visual defects in concrete floor slabs. In pavements, edges of cracks soon break off. They fill with debris, and, as concrete expands, material along the edges of the crack fails.

Cracks occur due to the volume change of concrete that is basically caused by drying shrinkage and temperature changes.

Control joints must be placed in slabs to allow for drying shrinkage. In floor slabs they should be spaced at intervals of not over 30 times the slab's thickness in both directions, resulting in slabs approximately square in shape.

Construction joints are usually located at the end of one day's pour or between lanes in pavement slabs. They are designed to allow load transfer with dowels or keyways.

Isolation joints are designed to separate slabs from such structures as column bases, catch basins, and walls. They are usually formed with a premolded filler, which is left in place.

Figure 7–22 illustrates types of joints for concrete construction.

7–8.6   Finishing concrete slabs involves screeding or striking off to the required grade, edging to form a slightly rounded corner at all joints, floating to remove any surface imperfections, and trowelling. The trowelling, usually with a power trowel, compacts the surface, interlocking the aggregate particles and the paste

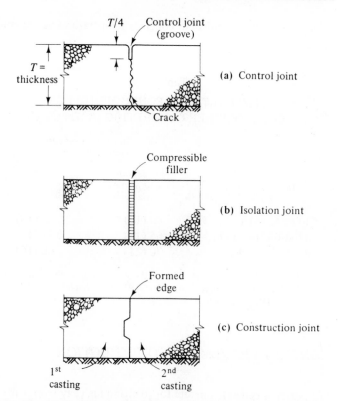

FIG. 7–22. Basic types of joints used in concrete construction. (Courtesy of the Canadian Portland Cement Association.)

in a dense hard layer. Finishing should not be attempted with any bleed water on the surface. This must be allowed to evaporate or removed with a squeegee before trowelling.

7–8.7 Special precautions and construction practices are required during very hot or very cold weather.

During hot weather, ice may be used to replace some of the mix water to keep the concrete from becoming too hot. Many specifications require that the temperature when placed be less than 29°C or 32°C (85°F or 90°F).

Combinations of high air temperature, high concrete temperature, low humidity, and high wind velocity may cause surface cracking. There is a risk of this when the rate of surface moisture evaporation exceeds 0.5 kg/m^2 per hour (0.1 lb/ft^2 per hour).

Cold weather concreting may require heating of the materials, enclosure of the site, and heating of the enclosed area. Concrete should not be allowed to freeze until it has gained sufficient strength to resist the disruptive forces.

Table 7–21 gives permissible concrete temperatures as recommended by ASTM.

Table 7–21

TEMPERATURE REQUIREMENTS FOR COLD WEATHER CONCRETING (ASTM C94)*

Air Temperature °F (°C)	Minimum Concrete Temperature °F (°C)	
	Thin Sections and Unformed Slabs	Heavy Sections and Mass Concrete
30 to 45 (-1 to 7)	60 (16)	50 (10)
0 to 30 (-18 to -1)	65 (18)	55 (13)
Below 0 (-18)	70 (21)	60 (16)

* Reprinted from ASTM C94 by the permission of the American Society for Testing and Materials, 1916 Race Street, Philadelphia PA 19103, Copyright.

7–9   CONCRETE PAVEMENT CONSTRUCTION

The use of portland cement concrete for road and highway pavements is growing with the advances in concrete technology and with concrete's increasing cost advantage with respect to petroleum products.

7–9.1   Slip form construction is now used for most concrete pavements. The base is prepared and the grade and alignment requirements laid out for the paving train. No slump concrete (slump less than 30 mm or 1 in) is placed on the roadbed in front of the train. The concrete is distributed, vibrated, screeded, and finished by the paving train as it proceeds. As the concrete is stiff enough to retain its shape shortly after it has been placed no side forms are required.

(The slip form method is also used for towers and silos. One example, the CN Tower in Toronto, was erected to a height of 460 m (1500 ft) in a little over 100 days using the slip form method.)

Tight quality control and consistent concrete are necessary for successful slip forming.

A modern concrete paving train operation with a slip form paver is illustrated in fig. 7–23.

7–9.2 Contraction and construction joints are usually required in pavement slabs.

Shrinkage cracking that would otherwise occur is controlled by one of three methods:

1. In unreinforced concrete, with contraction joints at 4–7 m (13–23 ft) spacing;
2. In lightly reinforced concrete, with contraction joints at greater intervals, in the range of 12–30 m (40–100 ft);
3. In heavily reinforced concrete, with no contraction joints.

Many fine hairline cracks are allowed to develop when the second or third method is used. They are prevented from opening to damaging widths by the reinforcement.

Maximum spacing for contraction joints in unreinforced slabs of the least of either 4.5 m or 30 times the thickness has also been recommended as well as longitudinal contraction joints between traffic lanes.

If contraction joints are utilized they are cut into the pavement with a concrete saw to a depth equal to one-quarter of the depth of the slab as soon as the concrete has stiffened sufficiently. Careful control of the timing of the cut is required. Concrete must be set hard enough so that the surface does not tear during cutting. However, waiting too long a time, one or two days, will result in the formation of random cracks.

Figure 7–24 shows a shrinkage crack in a pavement slab below the sawed joint.

The rough breakage of the crack below the sawed joint allows load transfer between slab segments.

Construction joints are also required between pours constructed at different times and at the edges, if the entire width is not constructed at one time. Dowells or keys are placed in these joints to act as load transfer devices.

Joints must be sealed to keep out water, dust, and other debris, and to ensure that movement of the slabs can continue to occur.

As is indicated in the next section, fiber reinforced concrete shows promise of decreasing the requirement for contraction joints in pavement slabs.

7–9.3 The pavement slab is often finished by a burlap drag, which results in a coarse sand-paper texture. The concrete is cured by wet burlap covers or by using a liquid membrane-forming sealing compound.

For pavements that will be subjected to the application of deicing chemicals, a period of at least one month for complete drying is required before exposure to these salts.

a

b

FIG. 7–23. Concrete pavers: (a) paving train on a highway project showing slip form paver and screeding, finishing and curing operations; (b) slip form paver, showing finished slab edge including a formed construction joint. (Courtesy of the CMI Corporation.)

FIG. 7–24. Contraction crack in a concrete pavement slab. The crack occurs under the sawed cut. (This is a construction joint showing the key formed in the side of the slab.)

7–10 SPECIAL CONCRETES

Many special-purpose concretes are available. New materials and construction methods are being developed. Design methods and quality control testing for these are beyond the scope of this book. However, some of these materials and construction techniques should be mentioned.

7–10.1 Pozzolanic materials are often used to produce portland-pozzolan cements, as admixtures in concrete, and to replace some of the portland cement in concrete.

Pozzolans are materials containing silica, with or without alumina, that react with calcium hydroxide to form compounds with cementitious properties. As indicated in section 7–1.2, the hydration of the cement chemicals produces the main cementing agent, calcium silicate hydrate, plus calcium hydroxide $[Ca(OH)_2)]$. About 20% of the reaction products are calcium hydroxide.

Fly ash, a residue from the combustion of pulverized coal, is a pozzolan that is readily available in many areas as refuse from coal-fired electric power plants.

In concrete, pozzolans increase workability and reduce possible damage due to alkali aggregate reaction and sulphate attack. Internal temperatures in mass concrete can be reduced with pozzolans.

The use of fly ash to replace some of the cement for economy results in lower early strength, but ultimate strength can be almost as high as with ordinary cement. Fly ash reacts slowly and proper curing conditions are important to ensure that the design properties are realized. Fly ash has been used to replace up to 35% of the portland cement in a mix without serious strength or other problems.

Ground blast-furnace slag is also readily available in many locations and has pozzolanic properties as well as cementitious properties of its own. Ground slag has also been used with small amounts of portland cement, aggregates, and water to produce an economical, flowing, fill material to use as pipe bedding, to refill utility cuts in roads, to form working slabs, for grouting, and other construction purposes.

Fly ash and ground slag are used extensively with normal, Type I (10) cement to produce cements for low heat or sulphate conditions, replacing Types II (20), IV (40) or V (50) portland cements at a more economical cost.

7–10.2 Lightweight concretes can be divided into two categories: structural lightweight concrete and insulating concrete.

Lightweight aggregates, expanded shale, slag, or certain types of volcanic rock are used.

Structural lightweight concretes have densities of 1400-1850 kg/m^3 (90-115 lb/ft^3) and strengths that are usually, but not necessarily, somewhat lower than normal concrete.

Insulating concretes have low strengths with densities as low as 240 kg/m^3 (15 lb/ft^3).

7–10.3 Heavyweight concretes are used for shielding purposes in the construction of nuclear reactors at power plants. Heavy aggregates, such as iron-ore particles, are used. Densities as high as 3200 kg/m^3 (200 lb/ft^3) have been obtained.

7–10.4 Architectural concrete, usually produced in precast plants, can be made with a wide variety of surface finishes, produced by using colors, exposing aggregates, sand blasting, brush hammering, and other techniques. Forms that produce special textures and rough surfaces are also used. Facing panels and other building components are manufactured.

7–10.5 Very high-strength concretes, 40-80 MPa (6000-12000 psi) or more, can be produced with today's technology. Superplasticizers allow plastic concrete with very low W/C ratios. Mechanical compaction increases strength greatly. Vacuum dewatering after the concrete has been placed but before it sets, can remove 20%-25% of the original water from the upper 100-200 mm (4-8 in) of a slab with dramatic improvements in strength and durability.

7–10.6 Fiber reinforced concrete contains about 2%-5%, by mass, short steel wires or fibers. Tensile strength of concrete is greatly increased. Thinner pavement slabs and greater distances between contraction joints are possible.

Asbestos, glass and polypropylene fibers have also been used to improve concrete's performance under tensile stresses.

7–10.7 "Rollcrete" or roller compacted concrete has been used in dam construction to place and strengthen concrete economically. A relatively low-cost mixture, with minimum water and cement content, can be placed using large economical earth moving equipment. Concrete is then compacted by rollers, resulting in much higher strengths than could be obtained with a similar mix without this type of compaction.

7–10.8 Polymer concrete or polymer-impregnated concrete contains organic materials that combine and grow into polymers, filling small pore space in concrete. In polymer concrete the material is added during mixing, whereas in the second type the concrete surface is coated and the liquid soaks into an existing slab.

These concretes have very high strengths and remarkable durability. Bridge decks, subjected to the application of deicing chemicals, have often deteriorated rapidly due to the corrosion of the reinforcing steel. Polymer impregnation has been used successfully to protect these surfaces by preventing the intrusion of the deicing chemicals.

7–10.9 Alumina cement is manufactured in the same manner as portland cement. However, the main product is calcium aluminate rather than calcium silicate. This cement hydrates rapidly, reaching high strengths in one day or less.

The rapid strength gain of alumina cement makes it useful in many construction or repair situations where time is important. It is also used in very cold climates where the large quantities of heat liberated and the rapid rate of hydration often allow construction to proceed without other protection.

Alumina cement is more resistant to most types of chemical attack than ordinary portland cement.

7–11   TEST PROCEDURES

Test procedures included are indicated below. Standard methods for these tests as well as many other tests on concrete can be found in the references shown in the Appendix.

7–11.1   Organic Impurities in Sand

7–11.2   Dry Rodded Density of Coarse Aggregates

7–11.3   Slump of Concrete

7–11.4   Air Content with the Pressure Meter

7–11.5   Casting Concrete Test Cylinders

7–11.6   Compressive Strength of Concrete Cylinders

7–11.7   Trial Mix–Water/Cement Ratio Method

7–11.8   Trial Mix–ACI Method

7–11.1   Organic Impurities in Sand

Purpose: To determine if there are organic compounds in sand that may be injurious to concrete.

Theory: Organic coatings on sand may retard setting of the concrete. The amount of these impurities can be checked by adding sodium hydroxide to the sample. The color of the sodium hydroxide solution changes, depending on the amount of organic material in the sand. A slight color change indicates that the amount of organics is not too injurious. However, if the color becomes dark amber, the sand should be rejected. A standard color chart is used to measure the color change. This contains five organic color plates: 1, 2, 3, 4, and 5. Color 3 is the dividing color.

Apparatus: 300-ml (12-oz) clear glass bottle
sodium hydroxide solution (3% by mass to 97% water)
color standard

Procedure:

1.   Fill the bottle to the 130-ml (4½-oz) mark with sand.
2.   Add the sodium hydroxide solution until the volume after shaking is 200 ml (7 oz).
3.   Shake vigorously; allow to stand for 24 hours.
4.   Compare the color of the liquid above the sand with the standard color plate.

Results: Record the color plate number that is closest to the color of the liquid in the bottle. If it is 1 or 2, the sand is acceptable; if it is 4 or 5, it is not; if it is 3, it is borderline.

7–11.2 Dry Rodded Density of Coarse Aggregates

Purpose: To obtain the approximate maximum dry density of coarse aggregate for concrete mixture.

Theory: The widely used ACI method of concrete design uses the dry rodded density of coarse aggregates. The most economical concrete mix contains the maximum amount of coarse aggregate particles. Therefore, the coarse aggregate is placed in a container of known volume, and then rodded and worked to get as many pieces as possible in the container; the resulting density is measured. If the concrete mix can be designed using this amount of coarse aggregate, the amounts of sand, cement, and water required to fill the spaces between the coarse aggregate particles will be a minimum.

Apparatus: Tamping rod
Containers of known volume (minimum) as follows:

Maximum Aggregate Size (in)	Capacity (ft^3)	Maximum Aggregate Size (mm)	Capacity (L)
½	1/10	20	7
1	1/3	50	15
1½	1/2	75	30
4	1		

Procedure:

1. Obtain a sample of clean, dry mixed aggregate.
2. Fill the container 1/3 full, level the surface, and rod 25 times.
3. Fill the container 2/3 full, level, and rod 25 times, rodding the second layer and just into first layer.
4. Fill the container, level, and rod 25 times as before.
5. Level the surface, and obtain the mass.

Results:

Volume of container _____
Net mass of aggregate _____

Calculations:

Dry rodded density _____

For saturated, surface-dry density, multiply by $1 + \dfrac{\% \text{ absorption}}{100}$.

7–11.3 Slump of Concrete

Purposes: To measure the slump or workability of concrete.

Theory: The amount that a sample of concrete, formed into a cone, slumps or falls is a measure of consistency or ease of placing and flowing.

Apparatus: slump cone
tamping rod
ruler

Procedure:

1. Dampen the cone and place it on a flat, moist surface.
2. With the operator standing on foot pieces, fill the cone in three layers with fresh concrete, rodding each layer 25 times. The rodding strokes should be uniformly distributed over the surface with a number near the perimeter. Rod the bottom layer through its depth and the second and third layers just into the layer below. Level off the top of the sample.
3. Lift the cone vertically until it is clear of concrete and set it down beside the sample. Place the tamping rod across the cone and over the concrete and measure the distance from the bottom of rod to the average height of the concrete.

Results: Measure and report the slump to nearest 10 mm or ¼ inch.

7–11.4 Air Content with the Pressure Meter

Purpose: To measure the air content of fresh concrete.

Theory: The pressure-type air meter has an air chamber on the lid. A hand pump allows this chamber to be pressurized. A bleed valve allows the pressure to be adjusted, and an operating valve allows air to enter the bowl containing the concrete. The operation of the meter is based on the principle that pressure times volume is constant. Assume the initial pressure in the chamber is P_1, the volume of air in the chamber is V_1, and the volume of air in the concrete is V_2. When the operating valve is opened, the air pressure in the chamber and the concrete (initially zero in the concrete) will become equal, P_2. As $P_1 V_1 = P_2 (V_1 + V_2)$, therefore, V_2 (volume of air in the concrete) $= V_1 (P_1 - P_2)/P_2$.

Values for P_1 and V_1 are constant, and the value for P_2 can be read from the pressure gauge, which is calibrated to read air content as a percentage directly.

Apparatus: pressure type air meter.

Procedure:

1. Fill the bowl of the air meter with fresh concrete in three layers. The concrete shall be consolidated in three layers by rodding each layer 25 strokes or vibrating each layer with a small vibrator inserted in three locations but not touching the bowl. Rodding shall be used for concrete with a slump of over 75 mm (3 in), vibrating for concrete with a slump of less than 40 mm (1½ in), and either method for concrete in between.
2. Level the surface of the concrete with the bowl surface, clean the lip, attach the lid, and clamp.
3. Fill the space between the bowl and the lid with water.
4. With a hand pump increase the air pressure in the air chamber to the initial point.
5. Open the operating valve, and read the air content.

Results: Record and report the air content as a percentage.

7–11.5   Casting Concrete Test Cylinders

Purpose: To cast concrete cylinders for subsequent compressive strength testing.

Theory: The strength of portland cement concrete is usually measured as the compressive strength of 150 mm × 300 mm (6 in × 12 in) cylinders of concrete, 28 days after casting, which have been cured in optimum conditions at 100% humidity and $23°C \pm 2°C$.

Apparatus: cylinder molds (metal or waxed cardboard)
              tamping rod or vibrators

Procedure:

1. Fill the cylinder in layers with fresh concrete, consolidating each layer with a tamping rod or a vibrator. Use a tamping rod from slumps over 75 mm (3 in), a vibrator for slumps under 40 mm (1½ in), and either for slumps in between. Rodding shall be 25 strokes per layer on each of three layers. With

vibration, the mold is filled in two layers, and the duration of vibration is limited to prevent segregation.
2. Level the surface of the mold with a trowel, cover it with a waterproof covering, and leave it for 20 hours or more without disturbance to set.
3. Remove the mold, and store the cylinders in a humid room or under water at $23°C \pm 2°C$ until time for the test.

Results: The cylinder number, the date of casting, and the type of consolidation should be recorded.

7–11.6   Compressive Strength of Concrete Cylinders

Purpose: To measure compressive strength of portland cement concrete.

Theory: The design strength of concrete is usually that reached by concrete cylinders at a certain age. To ensure that the true strength of the concrete in the cylinder is measured, it must be capped with a hard material to provide a flat, level loading surface with a plane perpendicular to the axis of the cylinder. A sulphur and fine sand mixture is usually used for this purpose. It is a liquid at $130°$-$145°C$ and hardens rapidly as it cools. To fail the cylinder, it must be loaded at a constant rate of between 150 and 350 kPa (20 to 40 psi) per second.

Apparatus: capping apparatus
compression test machine

Procedure:

1. Cap both ends of the cylinder in the capping mold with the sulphur compound.
2. Allow the cap to harden (should be 2–24 hours).
3. Place the cylinder in the compression machine and fail.

Results: Note the type of fracture—cone-shaped or other type of compressive fracture. The strength is the maximum load divided by the cross-sectional area (for standard 6-in cylinders, the area is 28.27 in^2 or 0.01824 m^2).

7–11.7   Trial Mix–Water/Cement Ratio Method

Purpose: To find the optimum proportions of cement, water, aggregates, and admixtures for a concrete mixture.

Apparatus: mixing pan
                  mixing tools
                  balances
                  slump cone
                  volumetric container
                  strength test molds
                  for air-entrained concrete a mechanical mixer must be used and an
                  air meter will be required.

Procedure:

1. Using given or established mix criteria, select the quantities required for a trial mix. Three kg (7 lb) of cement should be used for each standard cylinder to be cast. The water requirement can be obtained from the W/C ratio and the estimated amount of aggregates from fig. 7–16 and table 7–16.

2. If aggregates are not in the saturated, surface-dry condition, correct the calculated quantities for the actual water content.

3. Measure out the required quantities. For air-entrained concrete add the air-entraining agent to the water. For the aggregates measure out 20% more than the estimated requirement to allow for mix adjustments. Water can be measured by volume (1 ml = 1 g).

4. Dampen the mixing pan (or mixer), add the cement and water, and mix. Add 50%–75% of the aggregates and mix thoroughly. Additional fine and/or coarse aggregate should be added and mixed in until the workability and quality of the mix appear to be satisfactory. Record the amounts of each aggregate used, and note the quality of the mix.

5. Conduct a slump test and, if required, an air content test. (See sections 7–11.3 and 7–11.4.)

6. Remix the concrete; place it in a volumetric container in three layers, rodding each layer 25 times; level the surface flush with the top of the container; and obtain the mass of the concrete.

7. Remix the concrete and cast the strength test cylinders. (See section 7–11.5.)

8. Calculations:
   a. Find the density of the mix.
   b. Note the actual quantities used and find the adjusted quantities if the aggregates were not in the saturated, surface-dry condition.
   c. Calculate the mass per unit volume in lb/yd^3 or kg/m^3.

9. Cap and strength test the cylinders. (See section 7–11.6.)

CONCRETE TRIAL MIX–WATER/CEMENT RATIO METHOD
DATA SHEET

Design criteria: W/C _____ Slump _____ Max. agg. size_____

FM (fine agg.) _____ Air entrainment–yes _____ % or

no _____ (% entrapped = _____)

Batch quantities:

Cement _____ (A)

Water (W/C × cement) _____ (B)

Aggregate–total [factor (fig. 7–16) × water] _____

fine [% (table 7–16) × total] _____ %   _____ (C)

coarse (total – fine) _____ (D)

Air-entraining agent _____

Corrections for water content

	fine	*coarse*
water content, %	_____	_____
absorption, %	_____	_____
excess or required water, %	_____	_____
amount of excess water (% × mass of aggregate)	_____	_____
amount of required water (% × mass of aggregate)	_____	_____

Revised quantities

cement _____ (A)

water (B + required or – excess) _____ (B_1)

fine aggregate (C + excess or – required water) _____ (C_1)

coarse aggregate (D + excess or – required water) _____ (D_1)

air-entraining agent _____

Initial mass of aggregates measured out

fine (C_1 × 1.2) _____

coarse (D_1 × 1.2) _____

Mixture:

Aggregate used–fine _____ (C_2)

coarse _____ (D_2)

Appearance–stony, good, or sandy _____

Finishability–good, fair, or poor _____

Density:

Volume of container _____

Mass of concrete _____

Density _____ ρ

Slump: _____

Air content: _____

Calculations:
Materials used

cement	_____ (A)
water	_____ (B$_1$)
fine aggregate	_____ (C$_2$)
coarse aggregate	_____ (D$_2$)

Adjustments for water content

	fine	coarse
excess water in aggregate (% X mass used)	_____	_____
required water (% X mass used)	_____	_____

Adjusted quantities

cement	_____ (A)
water (B$_1$ + excess - required)	_____ (B$_2$)
fine aggregate (C$_2$ - excess or + required)	_____ (C$_3$)
coarse aggregate (D$_2$ - excess or + required)	_____ (D$_3$)
total	_____ (T)
air-entraining agent	_____

Mass per cubic meter

cement (A/T X ρ)	_____
water (B$_2$/T X ρ)	_____
fine aggregate (C$_3$/T X ρ)	_____
coarse aggregate (D$_3$/T X ρ)	_____

Strength test:

Age in days	_____
Failure load	_____
Area	_____
Strength	_____

7–11.8   Trial Mix–ACI Method

Purpose: To find the optimum proportions of cement, water, aggregates, and admixtures for a concrete mixture.

Apparatus: mixing pan and tools (or a mechanical mixer)
balances
slump cone
air meter
volumetric container
strength test molds

Procedure:

1. Using given or established mix criteria, find the estimated quantities required per cubic yard or cubic metre. The water requirement is obtained from table 7–18, the cement requirement from the water/cement ratio, and the coarse aggregate requirement from table 7–19. The amount of fine aggregate can be found by using the volume method or by using the estimated total mass of the concrete (table 7–20).

2. Calculate the amounts required for a trial mix. A batch of about 0.25 ft^3 or 0.007 m^3 should be mixed for each standard-strength cylinder to be made.

3. If the aggregates are not in the saturated, surface-dry condition, correct the calculated batch quantities.

4. Measure out the indicated amounts of each material, adding the air-entraining agent, if used, to the water.

5. Dampen the mixing pan (or mixer). Mix the cement and fine aggregates together. Add the coarse aggregate and mix. Add about 50%–75% of the water and mix thoroughly. Add additional water, mixing constantly, until the desired consistency is reached. (This may require more or less water than measured out.) Note the actual amount of water used and observe the quality of the mix.

6. Conduct a slump test and, if required, an air-content test. (See sections 7–11.3 and 7–11.4.)

7. Remix the concrete; place it in the volumetric container in three layers, rodding each layer 25 times; level the surface of the concrete flush with the top of the container; and obtain the mass of the concrete in the container.

8. Remix the concrete and cast the strength test cylinders. (See section 7–11.5.)

9. Calculations:
   a. Find the density of the concrete.
   b. Note the actual quantities used and calculate the adjusted amounts if the aggregates were not in the saturated, surface-dry condition.
   c. Calculate the mass per unit volume in lb/yd^3 or kg/m^3, and the actual W/C ratio.

10. Cap and strength test the cylinders at the specified age. (See section 7–11.6.)

Note: For a second trial mix, if required, use the yield quantities for proportions of water and fine and coarse aggregates, with any adjustments required to improve the workability or finishing properties. Recalculate the amount of cement needed, using the required W/C ratio and the indicated water content.

TRIAL MIX–ACI METHOD
DATA SHEET

Design criteria: W/C _____ Slump _____ Max. agg. size _____
Air entrainment–yes _____ % or no _____ (% entrapped = _____)
Fine aggregate–RD_{SSD} _____ FM _____
Coarse aggregate–RD_{SSD} _____ Dry rodded density _____
Estimated proportions–per cubic yard _____ or cubic metre _____

Water (table 7–18)   _____ (A)
Cement (water ÷ W/C)   _____ (B)
Coarse aggregate [volume (table 7–19)
  × dry rodded density]   _____ (C)
Air-entraining agent _____
Total mass   _____ (M)
Fine aggregate
(a) *Volume method*
  Volume water   _____
  Volume cement   _____
  Volume coarse aggregate   _____
  Volume air   _____
  Total   _____ (V)
  Volume fine aggregate
    (1 m^3 or 27 ft^3 – V)   _____
  Mass fine aggregate (volume × RD_{SSD}
    × ρ_W)   _____ (D)
(b) *Mass method*
  Estimated mass of concrete (table 7–20) _____ (M_1)
  Mass of fine aggregate (M_1 - M)   _____ (D)
Batch quantities–batch size   _____
  Water   _____ (A_1)
  Cement   _____ (B_1)
  Coarse aggregate   _____ (C_1)
  Fine aggregate   _____ (D_1)
  Air-entraining agent   _____
  Corrections for water content

	fine	*coarse*
water content, %	_____	_____
absorption, %	_____	_____
excess or required water, %	_____	_____
amount of excess water (% × mass of aggregate)	_____	_____
amount of required water (% × mass of aggregate)	_____	_____

Revised quantities
    water (A_1 – excess or + required)   _____ (A_2)
    cement   _____ (B_1)
    coarse aggregate (C_1 + excess or – required water)   _____ (C_2)
    fine aggregate (D_1 + excess or – required water)   _____ (D_2)

Mixture:
    Water used   _____ (A_3)
    Appearance—stony, good, or sandy   _____
    Finishability—good, fair, or poor   _____

Density:
    Volume of container   _____
    Mass of concrete   _____
    Density   _____ ρ

Slump: _____

Air content: _____

Calculations:
    Water used   _____ (A_3)
    Correction for aggregates
      + excess water   _____
      – required water   _____
    Mix water   _____ (A_4)
    Total mix ($A_4 + B_1 + C_1 + D_1$)   _____ (T)
    Mass per unit volume
      water ($A_4/T \times \rho$)   _____
      cement ($B_1/T \times \rho$)   _____
      coarse aggregate ($C_1/T \times \rho$)   _____
      fine aggregate ($D_1/T \times \rho$)   _____
    Actual W/C ratio (A_4/B_1)   _____

Strength test:
    Age in days   _____
    Failure load   _____
    Area   _____
    Strength   _____

7–12 PROBLEMS

Note: Problems 7–40 to 7–53 are in SI units and/or require the use of CSA specifications. Problems 7–54 to 7–67 are similar except for the use of traditional units and the reference to ASTM and ACI specifications.

7–1. What are the main compounds in portland cement? Which one is most important for early strength? For subsequent strength gain?

7–2. What is meant by hydration of portland cement? How does hydration affect the temperature of the material? How much water is required?

7–3. Name the types of portland cement. In what situation would each of these types of cement be used?

7–4. The strength of a normal concrete is found to be 24.7 MPa (3580 psi) at 28 days' age. Estimate the strength of this concrete at 6 months if it is moist cured.

7–5. What is cement clinker?

7–6. What are the main differences in chemical composition between Type I (10) and Type III (30) portland cements? Why are there these differences?

7–7. What is meant by water-cement ratio? What effect does the value of the water-cement ratio have on the strength of concrete? Why?

7–8. What is air-entrained concrete? When is it used? Why is it effective? What property of the mix governs the amount of air required? Why?

7–9. Estimate the percentage increase in 28-day strength for air-entrained concrete made with normal cement if the W/C ratio is reduced from 0.63 to 0.55.

7–10. What is meant by bleeding? Harshness?

7–11. The strength of a concrete is found to be 30 MPa (4350 psi) at 28 days' age. Estimate its strength at 3, 7 and 14 days of age.

7–12. Estimate the flexural strength of the concrete in problem 7–11 at 28 days' age.

7–13. Give three methods that can be used to increase concrete's resistance to disintegration from sulphate soil.

7–14. What is meant by segregation in concrete?

7–15. What are reactive aggregates? What precautions should be taken when this type of aggregate must be used in concrete?

7–16. Name four deleterious substances in aggregates for concrete and indicate why they are harmful.

7–17. Following are the results of a sieve analysis on a fine aggregate:

Passing 9.5 mm (3/8 in)	100.0%
4.75 mm (No. 4)	97.3
2.36 mm (No. 8)	82.8
1.18 mm (No. 16)	65.1
600 µm (No. 30)	48.6
300 µm (No. 50)	30.2
150 µm (No. 100)	12.9
75 µm (No. 200)	3.5

Find the fineness modulus for this aggregate.

7–18. Why are the amounts of material passing the No. 50 and No. 100 (300-μm and 150-μm) sieves important in concrete aggregates?

7–19. The W/C ratio required for an air-entrained concrete mix is 0.60. Estimate the amount of cement that could be saved per 100 cubic yards (76 cubic metres) of concrete if the allowable maximum aggregate size were increased from 3/4 in (19 mm) to 2 in (50 mm).

7–20. List four types of admixtures for concrete and indicate the purpose of each type.

7–21. When should air entrainment be used in concrete?

7–22. From fig. 7–12, obtain a W/C ratio that would produce air-entrained concrete, using normal cement, with a strength of 25 MPa or over in most mixes.

7–23. Using the water/cement ratio method, select quantities for a trial mix consisting of two strength cylinders to meet the following criteria: non-air-entrained; W/C ratio, 0.55; maximum aggregate size, 40 mm (1½ in); slump, 80–100 mm (3–4 in); fineness modulus of fine aggregate, 2.60.

7–24. Adjust the batch quantities found in problem 7–23 if the coarse aggregate is dry and the fine aggregate contains 4.5% water. Absorption is 1.2% and 0.7% respectively.

7–25. Using the ACI method, select batch quantities for a 0.80-ft^3 (0.0226-m^3) trial mix, with the following criteria: non-air-entrained; W/C, 0.58; maximum aggregate size, 3/4 in (20 mm); slump, 3–4 in (80–100 mm); coarse aggregate, RD_{SSD} = 2.69; dry rodded density, 104 lb/ft^3 (1670 kg/m^3); fine aggregate, RD_{SSD} = 2.66; and fineness modulus, 2.90. Use both volume and mass methods and compare the results.

7–26. Adjust the batch quantities found in problem 7–25 (volume method) if the condition of the aggregate is as described in problem 7–24.

7–27. What is meant by curing? What is required? Describe four methods used for curing concrete.

7–28. Describe three methods used to obtain higher than average strength in concrete at an early age (1–3 days).

7–29. What types of joints are used in concrete construction? Why is each used?

7–30. A concrete mix contains the following quantities per cubic yard:

Cement	516 lb
Water	273 lb
Fine aggregate	1355 lb
Coarse aggregate	1910 lb

Calculate the proportions in kg/m^3. Find the density in lb/ft^3 and kg/m^3.

7–31. Give two problems associated with over vibration of concrete.

7–32. Why are concrete floors trowelled?

7–33. What conditions may lead to surface cracking of concrete slabs in hot weather?

7–34. An unreinforced concrete sidewalk is constructed in the fall when the air temperature is 20°F (-7°C). What should the concrete temperature be according to the ASTM specifications?

7–35. What is "slip form" construction? What type of concrete is used?

7–36. An unreinforced pavement slab is 200 mm (8 in) thick. How far apart should the control joints be spaced?

7–37. What is fly ash? What type of material is it classified as? What chemical does it react with when hardening?

7–38. What is "fiber reinforced concrete"?

7–39. What type of cement would you recommend for patching on a highway if the road had to be reopened quickly?

7–40. A concrete mix contains 325 kg of cement and 170 kg of water per cubic metre. What is the W/C ratio?

7–41. Specifications require that a concrete mix contain at least 310 kg of cement per cubic metre and have a maximum W/C ratio of 0.55. How much water can be used in the mix?

7–42. A pavement slab 13 m in length was poured when the temperature was 10°C. Find the length of the slab after shrinkage (estimate the shrinkage as 500 μm/m for a dry mix) and when the temperature dropped to -22°C.

7–43. A floor slab, poured at a high water content, is 5 m by 5 m in size. Estimate the shrinkage.

7–44. Following are the results of a series of tests on a proposed fine aggregate for a pavement project. Check for acceptance according to the CSA specifications.

a. *Sieve analysis:*

retained on 10 mm (3/8 in)	0.0 g
5 mm (No. 4)	1.5
2.5 mm (No. 8)	7.3
1.25 mm (No. 16)	218.5
630 μm (No. 30)	85.5
315 μm (No. 50)	71.5
160 μm (No. 100)	59.1
Pan	19.0

b. *Organic content*—color, dark; corresponds to color plate 4.

c. *Washing over No. 200*—original sample   213.7 g
                        washed sample   211.1

d. *Soundness test with magnesium sulphate*— original sample   513.7 g
                                           final sample   421.3 g

7–45. Following are the results of tests on a concrete aggregate proposed for a concrete pavement subject to severe weathering. The aggregate is a crushed limestone, 20mm-5 mm (3/4 in to No. 4) in size. Check for acceptance by CSA requirements.

a. *Sieve analysis:*

retained on 28 mm (1 in)	0 g
20 mm (3/4 in)	375
14 mm (1/2 in)	1473
10 mm (3/8 in)	1161
5 mm (No. 4)	1796
2.5 mm (No. 8)	196
Pan	57

b. *Clay lumps*—sample size   203.5 g
            amount of clay lumps   5.1 g

c. *Washing over 80 μm (No. 200)*— original sample   1116.5 g
                          washed sample   1102.7 g

d. *Abrasion test*—original sample   5004 g
                after test   3162 g

e. *Soundness test with magnesium sulphate*—original sample   2015 g
                                              after test   1621 g

7–46. Give the CSA requirements for a concrete mix to be used for an unreinforced sidewalk slab to help ensure that the concrete will be durable.

7–47. Select trial mix criteria, using CSA requirements for:

a. A pavement, subjected to freezing and deicing chemicals, with a required strength of 25 MPa, 140 mm thick, without reinforcement.

b. A soil-retaining wall, 200 mm thick, exposed to freezing, reinforced with a 30-mm cover over the reinforcing bars, and with a required strength of 20 MPa.

c. A reinforced concrete floor slab, 150 mm thick, with reinforcing bars 100 mm apart and 40 mm from the surface, with a required strength of 20 MPa.

7–48. A concrete cylinder (150 mm) fails under a load of 298 kN. Find the compressive strength.

7–49. A flexural test beam 150 mm by 150 mm in size, fails at a load of 24.3 kN. Span length is 450 mm. Find the modulus of rupture.

7–50. Following are the approximate quantities required for 1 m³ of concrete. Check the volumes and adjust, if required, to produce 1 m³. [Relative density values (SSD) are 2.68 and 2.64 for the fine and coarse aggregate, respectively.]

Cement	330 kg
Water	195 kg
Fine aggregate	730 kg
Coarse aggregate	1250 kg
Air	3.5%

7–51. In a trial mix the following quantities of materials are used:

Cement	15.00 kg
Water	6.71 kg
Fine aggregate	28.42 kg
Coarse aggregate	36.01 kg

Density test results were—mass of concrete	16.27 kg
volume of container	0.25 ft³ (0.00708 m³)

The fine aggregate is dry (absorption is 1.0%) and the coarse aggregate contains excess water, a total of 4.2% (absorption is 1.2%). Adjust the mix quantities to reflect the water conditions of the aggregates, and calculate the proportions required for 1 m³.

7–52. A concrete plant uses the following proportions per cubic metre of concrete:

Cement	310 kg
Water	180 kg
Fine aggregate	750 kg
Coarse aggregate	1080 kg

The concrete is mixed in batches of 3.5 m³. Find the mass of each material required per batch if the fine aggregate conains 4.0% excess water. What will the W/C ratio be?

7–53. Design strength for the concrete on a project is 25 MPa. Results of strength tests on a series of 15 consecutive tests are as follows (MPa): 29.4, 26.3, 31.7, 28.8, 27.1, 30.4, 26.9, 29.1, 34.3, 24.1, 26.7, 23.3, 28.5, 32.3, and 27.6.

(a) Find the standard deviation for these results.

(b) Does this concrete meet the CSA requirements for strength? Why?

7–54. A concrete mix contains 550 lb of cement and 285 lb of water per yd^3.

(a) Find the W/C ratio by weight.

(b) Find the W/C ratio in gallons/sack.

7–55. Specifications require that a concrete mix contain at least 525 lb of cement per yd^3 and that the W/C ratio be 0.55 maximum. How much water can be used in the mix?

7–56. A sidewalk slab is 10 ft long. Estimate the drying shrinkage using average values Also estimate the change in length from summer (95°F) to winter (-18°F).

7–57. A floor slab, poured at a high water content, is 20 ft × 20 ft in size. Estimate the shrinkage.

7–58. Results of tests on a fine aggregate proposed are given in problem 7–44. Check these for acceptance according to the ASTM specifications.

7–59. Results of tests on a coarse aggregate are given in Problem 7–45. Check for acceptance according to the ASTM specifications.

7–60. A plain sidewalk slab is to be constructed in an area where deicing chemicals are often used. Give the ACI requirements for durability in this situation.

7–61. Select trial criteria, using ACI requirements, for:

(a) A pavement, subjected to freezing and deicing chemicals, with a required strength of 3500 lb/in^2, 7 in thick, without reinforcement.

(b) A soil-retaining wall, 8 in thick, exposed to freezing, reinforced with a 1-in cover over the reinforcing bars, and with a required strength of 3000 lb/in^2.

(c) A reinforced concrete floor slab, 6 in thick, with reinforcing bars 4 in apart and 1½ in from the surface, with required strength of 2500 lb/in^2.

7–62. A concrete cylinder (6 in) fails at 92000 lb. Find the compressive strength.

7–63. A flexural test beam, 6 in by 6 in in size, fails at a load of 5460 lb. Span length is 18 in. Find the modulus of rupture.

7–64. Following are the approximate quantities for 1 yd^3 of concrete.

Cement	515 lb
Water	305 lb
Fine aggregate	1120 lb
Coarse aggregate	1830 lb
Air	4%

Check the volumes of these constituents and adjust the quantities, if required, to produce 1 yd^3. G_{SSD} is 2.68 and 2.65 for the fine and the coarse aggregate respectively.

7–65. Using the data given in Problem 7–51, calculate the proportions required per cubic yard.

7–66. A concrete plant uses the following proportions per cubic yard of concrete:

Cement	510 lb
Water	210 lb
Fine aggregate	1180 lb
Coarse aggregate	1960 lb

Concrete is mixed in batches of 5.0 yd^3. Find the quantity of each material required per batch if the fine aggregate is wet and contains 4.5% excess water. What is the W/C ratio of the mixture?

7–67. Design strength for a concrete is 3500 psi. Results of strength tests on a series of 15 consecutive tests are as shown. The project is for ordinary concrete, not based on ultimate strength or for prestressing. Results are (psi) 4130, 3870, 4590, 4120, 3240, 3160, 3740, 3510, 3060, 3830, 4640, 4020, 3360, 3970, 4600.

(a) Find the standard deviation for these results.

(b) Does this concrete meet the ASTM specifications for strength?

Units and Symbols

The traditional North American system of units and the SI system are both used in this text. The SI system has been given preference as it will be gradually displacing the traditional system in engineering practice.

Basic SI units:

Length	metre	m	
Mass	kilogram	kg	
	gram	g	(1000 g = 1 kg)
	tonne	t	(1 t = 1000 kg)
Volume	litre	L	(1000 L = 1 m^3)
Force	Newton	N	(9.807 N = 1 kg force)
Pressure	Pascal	Pa	(1 Pa = 1 N/m^2)
Area	hectare	ha	(1 ha = 10 000 m^2)
Time	second	s	

SI prefixes:

Multiplication Factor	Prefix	Symbol
$1\ 000\ 000\ 000 = 10^9$	giga	G
$1\ 000\ 000 = 10^6$	mega	M
$1\ 000 = 10^3$	kilo	k
$0.01 = 10^{-2}$	centi	c
$0.001 = 10^{-3}$	milli	m
$0.000\ 001 = 10^{-6}$	micro	μ
$0.000\ 000\ 001 = 10^{-9}$	nano	n
$0.000\ 000\ 000\ 001 = 10^{-12}$	pico	p

Conversion factors (four significant figures):

Length	1 ft = 0.3048 m	Mass	1 lb = 0.4536 kg
	1 in. = 2.540 cm		1 ton = 0.9072 t
	1 mi = 1.609 km	Force	1 lb = 4.448 N
Area	$1\ ft^2 = 0.09290\ m^2$	Pressure	$1\ lb/in^2 = 6.895\ kPa$
	$1\ in^2 = 6.452\ cm^2$		$1\ lb/ft^2 = 47.88\ Pa$
	1 acre = 0.4047 ha	Viscosity	1 poise = 0.1 Pa·s
Volume	1 gallon (U.S.) = 3.785 L		$1\ stoke = 1\ cm^2/s$
	1 gallon (Can.) = 4.546 L	Density	$1\ lb/ft^3 = 16.02\ kg/m^3$
	$1\ in^3 = 16.39\ cm^3$		
	$1\ ft^3 = 0.02832\ m^3$	Concentration	$1\ lb/yd^3 = 0.5933\ kg/m^3$
	$1\ yd^3 = 0.7646\ m^3$		

Approximate relationships: These conversion factors are suitable for many calculations in soils and materials.

$62.4\ lb/ft^3 = 1000\ kg/m^3 = 1\ g/cm^3$ (density of water)
1 kg = 2.2 lb
1 m = 3.3 ft
$1\ m^3 = 1.3\ yd^3$
1 kg force = 9.8 N
1 km = 0.6 mi
$100\ kPa = 1\ ton/ft^2\ (1.04) = 1\ kg/cm^2\ (1.02)$
$= 15\ lb/in^2\ (14.5\ psi) = 1$ atmosphere

Symbols: The list that follows gives the symbols that are used in this text. Generally they are the symbols currently used in civil engineering practice in North America. Some symbols have been introduced, either to agree with the SI system, or to aid in the calculation of mass–volume relationships. The word *weight* and its symbol W have not been used to eliminate confusion between the use of weight to indicate either mass or force, as recommended in the SI system.

A	area	N	blow count from standard penetration test, normal force
AC	asphalt content		
AV	air voids		
c	cohesion	n	porosity
C_u	uniformity coefficient	NP	nonplastic
C_z	coefficient of curvature	Q	total flow, total load
C.B.R.	California Bearing Ratio	q	unit flow, unit load
d, D	diameter	q_q	quick shear compressive strength
D_{10}	diameter of the 10% finer size		
		q_u	unconfined compressive strength of soils
DTN	daily traffic number		
e	void ratio	RD	relative density (specific gravity)
EOS	equivalent opening size		
g	acceleration due to gravity	RD_A	apparent relative density
GI	Group Index	RD_B	bulk relative density
h, H	head, height of water	RD_{SSD}	saturated, surface-dry relative density
h_c	height of capillary rise		
i	hydraulic gradient	S	degree of saturation, shear force
I_D	density index		
I_P	index of plasticity	t, T	time
k	coefficient of permeability	V	volume
l, L	length	V_A	volume of air
M	mass	V_B	volume of asphalt (bitumen), bulk volume (aggregates)
M_A	mass of air		
M_B	mass of asphalt (bitumen)		
M_{BA}	mass of absorbed asphalt	V_{BA}	volume of absorbed asphalt
M_{BN}	mass of net asphalt	V_{BN}	volume of net asphalt
M_C	mass of cement	V_D	volume of dry soil
M_D	mass of dry soil	V_G	volume of aggregates
M_G	mass of aggregates	V_N	net volume
M_{SSD}	saturated, surface-dry mass	V_T	total volume
M_T	total mass	V_V	volume of voids
M_W	mass of water	V_W	volume of water
M_{WA}	mass of absorbed water	V_{WA}	volume of absorbed water

VMA	voids in mineral aggregate	ϵ	strain
w	water content	ϕ (phi)	angle of internal friction
w_L	liquid limit	ρ (rho)	density
w_o	optimum water content	ρ_D	dry density
w_P	plastic limit	ρ_W	density of water
w_S	shrinkage limit	σ (sigma)	normal stress
W/C	water–cement ratio	σ	standard deviation
ZAV	zero air voids	τ (tau)	shear stress

Appendix

Standard methods for conducting many of the tests outlined at the end of each chapter or referred to in the text can be found in the standards listed below. Many of the ASTM and the AASHTO standards are identical. The CSA standards included in A23.2, covering aggregates and concrete, are entirely in SI units.

Chapter	Test	ASTM	AASHTO	CSA
1	Relative Density of Soils	D854	T100	
	Grain Size Analysis of Soils	D422	T88	
	Atterberg Limits	D423,	T89,	
		D424	T90	
	Constant Head Permeability Test	D2434	T215	
	Unconfined Compressive Strength			
	of Soils	D2166	T208	
	Direct Shear Test of Soils	D3080		
	Undrained (quick) Shear Strength			
	in Triaxial Compression	D2850		
2	Soils Investigation by Auger Boring	D1452	T203	
	Thin-Walled Tube Sampling	D1587	T207	
	Penetration Test and Split Barrel			
	Sampling	D1586	T206	
	Field Vane Shear Test	D2573	T223	

3	Compaction Test—Standard	D698	T99	
	—Modified	D1557	T180	
	Field Density Test—Nuclear Method	D2922	T238	
	—Sand Cone Method	D1556	T191	
	—Balloon Method	D2167	T205	
4	Grain Size Analysis of Aggregates	C136	T27	A23.2-2A
	Amount Finer Than No. 200 Sieve	C117	T11	A23.2-5A
	Relative Density of Coarse Aggregate	C127	T85	A23.2-12A
	Relative Density of Fine Aggregate	C128	T84	A23.2-6A
	Abrasion Test—Los Angeles Apparatus	C131, C535	T96	A23.2-16A, -17A
	Soundness Test	C88	T104	A23.2-9A
	Petrographic Analysis of Aggregates	C295		A23.2-15A
	Clay Lumps and Friable Particles	C142	T112	A23.2-3A
	Lightweight Pieces in Aggregate	C123	T113	A23.2-4A
5	California Bearing Ratio	D1883	T193	
	Soil Cement—Making Test Specimens	D1632		
	—Compressive Strength	D1633		
	—Freezing and Thawing Tests	D560	T136	
	—Wetting and Drying Tests	D559	T135	
6	Ductility of Asphalt	D113	T51	
	Penetration of Asphalt	D5	T49	
	Density of Asphalt Mixtures	D1188	T166	
	Maximum Density of Asphalt Mixtures (Asphalt Absorption)	D2041	T209	
	Effect of Water on Strength of Asphalt Mixtures (Stripping)	D1075	T165	
	Coating and Stripping of Aggregates	D1664	T182	
	Asphalt Mix Design—Hveem Method	D1560	T246	
	—Marshall Method	D1559	T245	
	Extraction of Asphalt	D2172	T164	
7	Organic Impurities in Sand	C40	T21	A23.2-7A
	Reactivity of Concrete Aggregates	C227, C289, C586		A23.2-14A
	Density of Aggregates	C29	T19	A23.2-10A
	Slump of Concrete	C143	T119	A23.2-5C
	Air Content of Concrete			
	—Pressure Method	C231	T152	A23.2-4C
	—Volume Method	C173	T196	A23.2-7C

Making and Curing Concrete Test Specimens	C192	T126	A23.2-3C
Capping Concrete Cylinders	C617	T231	
Compressive Strength of Concrete Specimens	C39	T22	A23.2-9C
Flexural Strength of Concrete Specimens	C78	T97	A23.2-8C
Accelerated Curing of Concrete Specimens	C684		A23.2-10C
Density and Yield of Concrete	C138	T121	A23.2-6C

Bibliography

The American Association of State Highway and Transportation Officials. *AASHTO Interim Guide for Design of Pavement Structures, 1972.* Washington, D.C.: The Association, 1974.

———. *AASHTO Materials, Part I, Specifications,* and *Part II, Tests.* Washington, D.C.: The Association, 1978.

The American Society for Testing and Materials. *Annual Book of ASTM Standards, Part 14, Concrete and Mineral Aggregates.* . . . Philadelphia: The Society, 1977.

———. *Annual Book of ASTM Standards, Part 15, Road and Paving Materials.* . . . Philadelphia: The Society, 1977.

———. *Annual Book of ASTM Standards, Part 19, Soil and Rock.* . . . Philadelphia: The Society, 1977.

The Asphalt Institute. *Asphalt Hot-Mix Recycling,* first ed. Manual Series No. 20 (MS-20). College Park, Md.: The Institute, 1981.

———. *The Asphalt Handbook.* Manual Series No. 4 (MS-4). College Park, Md.: The Institute, April 1965.

———. *A Brief Introduction to Asphalt,* 7th ed. Manual Series No. 5 (MS-5). College Park, Md.: The Institute, September 1974.

———. *Mix Design Methods for Asphalt Concrete and Other Hot-mix Types,* 4th ed. Manual Series No. 2 (MS-2). College Park, Md.: The Institute, 1974.

Thickness Design—Asphalt Pavements for Highways and Streets. Manual Series No. 1 (MS-1). College Park, Md.: The Institute, 1981.

Baker, Robert F.; Byrd, L. G.; and Mickle, D. Grant, eds. *Handbook of Highway Engineering.* New York, Toronto: Van Nostrand Reinhold Co., 1975.

Bowles, Joseph E. *Engineering Properties of Soils and Their Measurement,* 2nd ed. New York, Toronto: McGraw-Hill Book Co., 1978.

The Canadian Portland Cement Association. *Design and Control of Concrete Mixtures,* Metric ed. Toronto: The Association, 1978.

Canadian Standards Association. *Concrete Materials and Methods of Concrete Construction–Methods of Test for Concrete.* Rexdale, Ontario, The Association, 1977.

Craig, R. F. *Soil Mechanics.* New York: Van Nostrand Reinhold Company, 1974.

Herubin, Charles A., and Marotta, Theodore W. *Basic Construction Materials.* Reston, Va: Reston Publishing Co., Inc., 1977.

Hough, B. K. *Basic Soils Engineering,* 2nd ed. New York: Ronald Press Co., 1969.

Jackson, N., ed. *Civil Engineering Materials.* London: Macmillan Press Ltd., 1976.

Jumikis, Alfreds R. *Introduction to Soil Mechanics.* Princeton, Toronto: D. Van Nostrand Co., Inc., 1967.

Krebs, Robert D., and Walker, Richard D. *Highway Materials.* New York, Toronto: McGraw-Hill Book Co., Inc., 1971.

Lambe, William T., and Whitman, Robert V. *Soil Mechanics.* New York, Toronto: John Wiley and Sons, Inc., 1969.

Larson, Thomas D. *Portland Cement and Asphalt Concretes.* New York, Toronto: McGraw-Hill Book Co., Inc., 1963.

Malhotra, V. M., ed. *Progress in Concrete Technology.* Ottawa: Energy, Mines, and Resources, 1980.

McCarthy, David F. *Essentials of Soil Mechanics and Foundations,* Reston, Va.: Reston Publishing Co., Inc., 1977.

Means, R. E., and Parcher, J. V. *Physical Properties of Soils.* Columbus, Ohio: Charles E. Merrill Books, Inc., 1973.

Mindess, Sidney and Young, J. Francis. *Concrete.* Englewood Cliffs, NJ: Prentice-Hall, Inc., 1981.

Murdock, L. J., and Blackledge, G. F. *Concrete Materials and Practice,* 4th ed. London: Edward Arnold Ltd., 1968.

Oglesby, Clarkson H. *Highway Engineering,* 3rd ed. New York, Toronto: John Wiley and Sons, 1975.

Patton, W. J. *Construction Materials.* Englewood Cliffs, N.J.: Prentice-Hall, Inc., 1976.

Peck, Ralph B.; Hanson, Walter B.; and Thornburn, Thomas H. *Foundation Engineering,* 2nd ed. New York, Toronto: John Wiley and Sons Inc., 1975.

Portland Cement Association. *Design of Concrete Pavement for City Streets.* Chicago: The Association, 1974.

Portland Cement Association. *Soil Cement Construction Handbook.* Chicago: The Association, 1969.

———. *Soil Cement Laboratory Handbook.* Chicago: The Association, 1971.

———. *Thickness Design for Concrete Pavements.* Chicago: The Association, 1966.

Ritter, Leo J., Jr., and Paquette, Radnor J. *Highway Engineering,* 3rd ed. New York: Ronald Press Company, 1967.

Sargious, Michel. *Pavements and Surfacings for Highways and Airports.* New York, Toronto: Halsted Press—John Wiley and Sons, 1975.

Schroeder, W. L. *Soils in Construction.* New York, Toronto: John Wiley and Sons, Inc., 1975.

Sowers, George B., and Sowers, George F. *Introductory Soil Mechanics and Foundations,* 3rd ed. London: Macmillan Company, 1970.

Spangler, Merlin G., and Handy, Richard L. *Soil Engineering,* 3rd ed. New York, London: Intext Educational Publishers, 1973.

Teng, Wayne C. *Foundation Design.* Englewood Cliffs, N.J.: Prentice-Hall, Inc., 1962.

Troxel, George Earl, and Davis, Harmen E. *Composition and Properties of Concrete.* New York, Toronto: McGraw-Hill Book Co., Inc., 1956.

Wallace, Hugh A., and Martin, J. Rogers, *Asphalt Pavement Engineering.* New York: McGraw-Hill Book Co., 1967.

Woods, Kenneth B.; Berry, Donald S.; and Goetz, William H. *Highway Engineering Handbook.* New York, Toronto: McGraw-Hill Book Co., Inc., 1960.

Wu, Tien Hsing. *Soil Mechanics.* Boston, London, Sydney: Allyn and Bacon, Inc., 1976.

Yoder, E. J., and Witczak, M. W. *Principles of Pavement Design,* 2nd ed. New York, Toronto: John Wiley and Sons, Inc., 1975.

Index